EMERGING LOCATION AWARE BROADBAND WIRELESS AD HOC NETWORKS

T0138185

EMERGING LOCATION AWARE BROADBAND WIRELESS AD HOC NETWORKS

Edited by

Rajamani Ganesh
QUALCOMM, India

Sastri L. Kota
HARRIS CORPORATION, USA

Kaveh Pahlavan
WORCESTER POLYTECHNIC INSTITUTE, USA

Ramón Agustí
UNIVERSITAT POLITÈCNICA DE CATALUNYA, Spain

Springer

eBook ISBN: 0-387-23072-6
Print ISBN: 1-4899-8342-2

©2005 Springer Science + Business Media, Inc.

Print ©2005 Springer Science + Business Media, Inc.
Boston

Created in the United States of America

Visit Springer's eBookstore at: http://ebooks.kluweronline.com
and the Springer Global Website Online at: http://www.springeronline.com

Contents

PART IV   ENCODING, ALGORITHMS AND PERFORMANCE

Preface

Wireless networking has emerged as one of the major areas for research and industrial development in the past couple of decades and is expected to continually expand into new areas. Extension of the cellular voice telephony to adhoc and mesh networks, wireless hotspot Internet access, wireless home and corporate networking, and wireless personal area networks have profoundly impacted our lifestyle. After more than a decade of exponential growth, today's wireless industry is one of the largest industries in the world. At the time of this writing, more than one billion people subscribe to cellular telephone services worldwide, billions of short messages are weekly exchanged, wireless local area networks are connecting home computers and provide access in corporate and public hotspots and yet we are expecting emerging adhoc networks to connect a myriad of terminals that range from small sensors to huge appliances in the near future. The penetration of the cellular telephone in some countries such as Luxemburg has exceeded the population of the country and there are over 250 million subscribers in China alone. In response to this growth, a number of engineers in the industry and students in the academia are eager to update their knowledge in this field of wireless networks. These people need books that provide an overview of the existing issues in the wireless network industry from different points of view. This incentive has encouraged the editors of this book to prepare a collection of the views, addressing the technical challenges, possible solutions, and experimental results, from a number of diverse resources in a timely manner.

This book on Emerging Local Aware Broadband Wireless Adhoc Networks provides a collection of 19 invited papers for the International Symposium on Personal, Indoor, Mobile, Radio Communications (PIMRC) held in Barcelona, Spain in September 5-8, 2004. Started at the King's College, University of London in 1989, PIMRC is one of the leading international conferences in wireless networking that has been frequently held in different countries in the Europe, North America and Far-East. Today, the PIMRC conference enjoys wide respect and represents a new trend in international conferences that is well suited to the evolving global information technology market. PIMRC is uniquely identified by its balance among academia, industry, and governmental organizations. Its international technical program has been endorsed by IEEE of the US, IEE of the UK, and IEICE of Japan. In the past, a number of top executives of the major manufacturers of wireless equipment and wireless service providers, minister of telecommunications and other top executives of government organizations, and world-renowned research scientists in wireless networks have participated in this annual event.

To enhance the overall technical depth of the presented material at the PIMRC 2004, nineteen distinguished contributors in this field were invited to present us with a written text of their presentation material that is collected in this book edition. The invited papers are organized into four parts:

Part I   Trends in Wireless Networks
Part II   Wireless LANs and Adhoc Networks
Part III   Mobile Wireless Internet and Satellite Applications
Part IV   Encoding, Algorithms and Performance

Part I, Trends in wireless networks covers six chapters. In Chapter one a multilayer system model of the WCDMA/3G to understand the inter layer protocol interactions is proposed. TCP performance analysis under joint rate and power adaptation with constrained BER requirements for down link data transmission in a cellular Variable Spreading Factor (VSF) WCDMA/3G network is included. A general model describing the interaction between TCP, radio link adaptation and the error control protocol is presented. As the services like push to talk becomes popular, there is a definite need for "always on services". Chapter 2 defines requirements for such a push to talk service and describes an overlay network based solution with initial prototype experimental results. As new trends are emerging, several hard problems with potential high payout warrant research solutions. In Chapter 3 some of the topics shared among the fixed wired and wireless network domains are discussed with possible approaches to lower the complexity of the communication systems. The use of multiple antenna systems e.g. smart antennas with adaptive beam forming and MIMO systems improves the spectrum efficiency and achieving high data rate within wireless systems.

Chapter 4 describes such multiple antenna systems applicable to the third generation including Universal Mobile Telecommunication System (UMTS), the wireless LAN, wireless PAN and up to fourth generation wireless communication systems. Chapter 5 is devoted to a fixed relay station acting as a wireless bridge concept along with performance evaluation for a wireless broadband system considering a Manhattan-like dense urban traffic environment. Chapter 6 revisits cellular architecture as applied to the CDMA air interface and the paradigm changes of the service from coverage limited mobile traffic to a mix of urban pedestrian, in-building usage and from voice-only to a mix of data services. The salient features of CDMA, service optimization with modification of air access topology and control are discussed with relevant examples.

Part II, Wireless LANs and Adhoc Networks consists of five chapters describing the Voice over IP application carried over wireless Local area Network (LAN) and adhoc network architectures and performance. In chapter 7, a novel IEEE 802.11 based self-organizing hierarchical adhoc wireless network (SOHAN) consisting of three tier low power mobile nodes, high power forwarding nodes and wired access points without power constraints has been proposed. Initial experimental results along with hardware and software implementation requirements are described. Chapter 8 presents experimental results for the capacity of Voice over IP (VoIP) in an IEEE 802.11b network and quantifies contributing factors that limit the capacity to only less than 10 voice calls per access point. Chapter 9 introduces an adhoc wireless mobile network architecture employing a hierarchical notion. A Mobile Backbone Network Protocol (MBNP) covering proactive routing schemes, Media Access Control (MAC) layer power control, flow admission control mechanisms and network/MAC cross layer resource allocation schemes is proposed and described. Various design issues and the Quality of Service (QoS) guarantees are included. Compared with a cellular network an adhoc network is less efficient due to lack of stationary infrastructure. Chapter 10 proposes a novel Vertex-Linked Infrastructure (VLI) for adhoc network architecture. This approach uses an easily deployable, survivable, wired infrastructure as a backbone of the adhoc network. Chapter 11 surveys probabilistic models for positioning problem and related issues like calibration, active learning, error estimation and tracking with history.

Part III dealing with Mobile Wireless Internet and Satellite Applications consists of five chapters. In order to address the uncertainty associated with wireless mobile networks achieving seamless integration with Internet, delivering the required QoS warrants new protocol designs and implementations. Chapter 12 proposes an overarching theoretical framework for representing relevant network information in terms of underlying entropies, entropy rates and their interrelationships. Application of the proposed framework to design optimal mobility tracking and resource management coping with the uncertainty in traffic load, topology control and routing is illustrated. Chapter 13 addresses Internet mobility allowing a computer to move from network to network, and provider to provider without disrupting its active sessions. The proposed methodology is based on the marriage of cellular mobility techniques with the Internet protocol suite. A notional sketch of the resultant IPv6 mobility architecture is introduced. Chapter 14 covers the performance issue of mobile Internet consisting of several wireless networks with limited bandwidth and higher error rates than the wired links. An overview of a number of proxy services including header/data compression for wireless application protocol, vertical handover support between cellular and wireless local area networks, and multiple-channel transport layer security is discussed. Next two chapters address trends, developments and applications of Satellite communications. Chapter 15 describes important satellite communications technology developments such as digital video broadcasting, on-board switching with QoS support, traffic management, and satellite multiprotocol label switching (MPLS). Seamless integration of satellite network with terrestrial wireless access networks and satellite communications for aeronautical applications are also discussed. Chapter 16 addresses the question of whether satellites have a role in mobile communications other than for niche areas such as sea and aero coverage. History of mobile satellite communication systems and current developments towards multimedia, broadband, content delivery and integrated systems is described. An illustration through B3G era on how satellites can play a useful role in cooperation with terrestrial systems is described.

Finally, Part IV deals with Encoding trends, Algorithms and Performance models and results in three chapters. Chapter 17 reviews the progress of channel coding techniques starting from turbo codes and the rediscovery of the low-density parity check (LDPC) codes which essentially closed the gap between the Shannon Capacity limit and practical implementation. Application of turbo codes and LDPC codes to the 3G wireless standards e.g. 3GPP and 3GPP2 and digital video broadcast for satellite standard (DVB-S2) are illustrated. Chapter 18 proposes a network wide Time-Diffusion Synchronization Protocol (TDP) for future sensor networks consisting of small intelligent devices deployed in homes, plantations, oceans, streets and highways. Simulation results are included to validate the effectiveness of TDP in synchronizing the time through out the network and balancing the energy consumed by the sensor nodes. In Chapter 19 the choice of basis for linear matrix modulation (linear space-time code with linear combination alphabet) is considered. Unitary invariants are discussed with a full spectrum of invariants between the well-known trace and determinant. Examples of symbol rate 3 schemes for 4 transmit antennas are discussed.

The editors of this book would like to express their appreciation to all the invited authors for their excellent contributions meeting a very stringent time schedule. We would like to thank the Technical Program Committee, in particular, Prof. Oriol Sallent, Co-Chair of the PIMRC 2004 for the encouragement in preparation of this book. Our appreciation is also due to Mr. Alex Greene, Ms Melissa Sullivan and others at Kluwer Academic Publishers for their cooperation and help in meeting the publication dates of the PIMRC 2004.

Rajamani Ganesh

Sastri L. Kota

Kaveh Pahlavan

Ramon Agusti

with a full spectrum of invariants between the well-known trace and determinant. Examples of symbol rate 3 schemes for 4 transmit antennas are discussed.

The editors of this book would like to express their appreciation to all the invited authors for their excellent contributions meeting a very stringent time schedule. We would like to thank the Technical Program Committee, in particular, Prof. Oriol Sallent, Co-Chair of the PIMRC 2004, for the encouragement in preparation of this book. Our appreciation is also due to Ms. Alex Greene, Ms. Melissa Sullivan and others at Kluwer Academic Publishers for their cooperation and help in meeting the publication dates of the PIMRC 2004.

Ramjee Prasad

Sastri L. Kota

Kaveh Pahlavan

Ramon Agusti

PART I

TRENDS IN WIRELESS NETWORKS

Chapter 1

CROSS-LAYER PERFORMANCE IN CELLULAR WCDMA/3G NETWORKS: MODELLING AND ANALYSIS

EKRAM HOSSAIN[1], VIJAY K. BHARGAVA[2]

[1]*Department of Electrical and Computer Engineering, University of Manitoba, Winnipeg, MB, Canada R3T 5V6, Tel: (204) 474 8908, Fax: (204) 261 4639, Email: ekram@ee.umanitoba.ca;* [2]*Department of Electrical and Computer Engineering, University of British Columbia, 2356 Main Mall, Vancouver, BC, Canada V6T 1Z4, Tel: (604) 822 2342, Fax: (604) 822 0604, Email : vijayb@ece.ubc.ca*

Abstract: Dynamic radio link adaptation is a key component in WCDMA/3G wireless networks to improve the spectral efficiency while meeting the radio link level QoS (Quality of Service) requirements such as the BER (Bit Error Rate) requirements for the different wireless services. Again, performance of an end-to-end protocol, such as TCP (Transmission Control Protocol) depends on the performance of the underlying radio link adaptation technique. A multilayer modelling of the WCDMA/3G radio interface is therefore necessary to better understand the inter-layer protocol interactions and identify suitable transport and radio link layer mechanisms to improve TCP performance in a wide-area cellular WCDMA/3G network. In this article, we present such a multilayer system model to analyze TCP performance under joint rate and power adaptation with constrained BER requirements for downlink data transmission in a cellular VSF (Variable Spreading Factor) WCDMA/3G network. To this end, we present the outline of a more general model which can completely describe the interaction between TCP and the radio link adaptation and error control protocol. We give an example of using such a model to design radio link layer scheduler for achieving high wireless link utilization.

1.   INTRODUCTION

WCDMA/3G networks are expected to provide high speed packet data services including wireless Internet access [1]. Since the transport layer protocol performance is one of the most critical issues in data networking over noisy wireless links, the performance of TCP, which is the flagship protocol in today's Internet, would be crucial in such an environment [2]. TCP is a connection-oriented transport layer protocol which guarantees reliable, in-sequence delivery of packets and is generally more suited for delay-insensitive applications.

The performance of TCP in a wireless network depends on the service provided by the underlying RLC (Radio Link Control)/MAC (Medium Access Control) protocol. In a WCDMA/3G system, the transmission rate and the power corresponding to the different mobile users can be dynamically varied depending on the variations in channel interference and fading conditions to improve the wireless channel utilization while meeting the lower layer (e.g., RLC/MAC layer, PHY layer) QoS (e.g., BER) requirements [3]. This is referred to as the dynamic link adaptation in this article.

Dynamic link adaptation for optimizing the radio link level performance and the issue of improving TCP performance in wireless networks have often been addressed in literature as separate problems, especially for CDMA networks. Transmission protocol stack performance in a cellular wireless network would be optimized if the radio link/PHY level and the transport level protocol performances are jointly optimized, and a multilayer modelling is therefore necessary to explore the issues of inter-layer protocol interaction (e.g., implications of link adaptation on higher layer protocol performance) and identify suitable radio link and transport protocol mechanisms. A radio link level protocol based on dynamic channel coding with retransmission was proposed in [4]. Radio link level techniques to improve TCP performance in narrowband wireless systems were proposed in [5]-[9]. Effects of link layer retransmissions on TCP performance were investigated in [10] and [11]. Some of the earlier works on wireless TCP (e.g., [12]) investigated the performance of a single circuit-switched TCP connection in a constant spreading factor DS-CDMA network. In [13], the throughput performance of TCP over 3G wireless links was analyzed in the presence of varying bandwidth and delay. Throughput and energy performances of circuit-switched TCP connections in a wideband CDMA air interface in the absence of link level retransmissions were investigated in [14].

The system dynamics representing the interactions between the higher layer and the lower layer protocols in an interference-limited cellular WCDMA environment is fairly complex because of the existence of numerous parameters and non-linear nature of the protocol state machines at the different layers. For a particular system load, radio link level rate adaptation and scheduling, persistence in the retransmission of the failed radio frames, the desired signal-to-interference ratio and the physical level channel coding mechanisms impact the radio link level and hence the higher layer protocol performance. Again, under a particular radio link level configuration, transport layer mechanisms such as the congestion and flow control and adaptive retransmission timeout determine the achieved transport layer protocol throughput.

In this article, we evaluate some aspects of TCP performance for *downlink* (*i.e., forward link*) data transmission in a cellular WCDMA/3G network where the dynamic link adaptation is assumed to be achieved by joint rate and power adaptation under constrained BER requirements. Variable rate transmission is implemented by using the variable spreading factor (VSF) method. Two joint rate and power allocation algorithms, namely, *Depth-First Allocation* (*DFA*) and *Breadth-First Allocation* (*BFA*), are considered for dynamically adjusting the transmission rate and power corresponding to the different TCP connections. The DFA and the BFA algorithms use exhaustive and round-robin principles, respectively, for dynamic resource allocation. To this end, we introduce a novel model to describe completely the interaction between TCP and the radio link control protocol in a more general set up. Such a model would enable us to design cross-layer optimized radio link control protocol for future-generation wireless networks. We present an example where the transport-layer information is exploited to design a *cross-layer scheduler* at the radio link level to improve TCP throughput performance (and hence wireless channel utilization).

2. A MULTILAYER SYSTEM MODEL FOR CELLULAR WCDMA/3G NETWORKS

2.1 TCP Model

A sender-based one way traffic scenario is considered where the mobile stations (MSs) act as TCP sinks. To study the steady-state behavior of TCP, bulk data TCP connections are assumed (i.e., senders always have data to transmit and can transmit as many packets as their transmission windows allow). The TCP sinks can accept packets out of sequence but deliver them only in sequence to the user and they generate immediate acknowledgments (ACKs).

Similar to that in all currently available TCP implementations, TCP congestion and error control algorithms include slow-start, congestion avoidance, fast retransmit and fast recovery [2]. The RTO (Retransmission Time Out) value at the sending TCP host is updated based on the recommendations in [15].

2.2   Radio Link Layer Model

A transport protocol data unit is typically forwarded to the radio link level protocol (with the necessary intermediate protocol overheads appended to it), where it is segmented into several radio frames for transmission in the air interface. The radio link control protocol for downlink transmission to the mobiles has two components - a mechanism for dynamic rate allocation (i.e., selection of the mobile(s) and the number of radio frames to be transmitted to the corresponding mobile(s) during a frame-time) and a mechanism for error control.

2.2.1 Dynamic Rate Allocation/Link Adaptation

With variable rate frame transmission, the number of radio frames transmitted during a frame-time varies according to the transmission rate used during that frame-time. Let us assume that the transmission rates can be selected from the set of rates $\{r_0, r_1, r_2, ..., r_\varphi\}$ and $r_m = mr_1$ ($m = 0, 1, ..., \varphi$) so that the normalized value of r_m with respect to the *basic* rate r_1 is m. Therefore, if the frame length corresponding to the basic rate r_1 is M bits, the frame length for rate r_m is mM bits. For this, a variable spreading factor WCDMA system can be used where the basic gain is given by N chips per bit, and for rate r_m, the spreading gain is reduced to N/m chips per bit.

The radio link level transmission rate corresponding to a TCP connection is determined by a dynamic rate adaptation algorithm and is based on the system load (in terms of the number of concurrent TCP connections) and the channel interference and fading conditions corresponding to the mobile TCP sinks. We consider the two following dynamic rate allocation mechanisms which were developed based on a general SIR model in a cellular VSF WCDMA network using the concept of *rate capacity resource* in a cell [3].

<u>Depth-First Allocation (DFA)</u>
In this case, the maximum possible transmission rate is allocated to the connection with the lowest interference factor and then if rate capacity resource is available, the maximum possible transmission rate is allocated to the connection with the next lowest interference factor. In this way, transmission rates are assigned to the connections in a 'depth-first' fashion

starting from the one for which the corresponding interference factor is the lowest. Although this 'greedy' mechanism will maximize the radio link level throughput, the throughput fairness may degrade.

Breadth-First Allocation (BFA)

In this case, the radio link level transmission rate corresponding to each TCP connection is first set to be inversely proportional to the corresponding interference factor. Then the rate allocations corresponding to all TCP connections are incremented in a round-robin fashion as far as the sum interference is less than the rate capacity resource [3].

2.2.2 Radio Link Level Error Control

We consider an SR (Selective Repeat)-hybrid-ARQ protocol enhanced for radio link level variable rate transmission. In this error control method forward error correction coding is applied to each of the radio frames transmitted during a frame-time so that each of them can be decoded individually at the mobile receiver [1]. Only the frames transmitted in error are retransmitted.

2.2.3 Scheduling and Buffering

Per-destination queuing with FIFO (First-In-First-Out) scheduling (within each queue) is assumed for the TCP segments at the BS. The TCP segments corresponding to the different connections are assumed to arrive at the BS in their original transmission order. The buffer size at the BS is assumed to be sufficiently large so that there is no buffer overflow loss. The radio link level frames corresponding to a TCP segment are buffered in the corresponding RLC/MAC level queue. From the ith RLC/MAC layer queue, m_i frames are transmitted during each frame-time, where m_i ($0 \le m_i \le \varphi$) is determined using the dynamic rate adaptation procedure.

2.3　Channel Fading and Micro-Mobility Model

The radio link qualities of the mobile users depend on the spatial distribution (and hence the micro-mobility patterns) of the users along with the short-term and the long-term channel fading conditions. We consider a random micro-mobility model [16]. The nature of the long-term fading (i.e., shadowing) is assumed to be uncorrelated. Since the short-term fading changes more rapidly compared to the long-term fading, it is assumed to be independent over successive link adaptation intervals.

3.   PERFORMANCE EVALUATION

3.1 Performance Metrics and Simulation Parameters

Two performance metrics are considered here–*average per-connection TCP throughput (λ)* and TCP *throughput fairness (F)*. TCP throughput for the *i*th TCP connection in cell *j* ($\lambda_i^{(j)}$) is measured as the amount of successfully transmitted TCP data per second for this connection.

The TCP throughput fairness F among $\lambda_i^{(j)}$ for all the TCP connections measures the global fairness in average TCP throughput [16] and $F = 1.0$ corresponds to the ideal case of perfect throughput fairness. Note that, even for TCP connections with similar round-trip time throughput unfairness may be induced by wireless channel errors and the dynamics of the radio link level variable rate allocation.

Simulation results are obtained under varying traffic load (i.e., number of concurrent TCP connections) distribution across a 3-cell system in a hexagonal cell-layout. With one bulk-data TCP connection per MS, the average number of concurrent TCP connections per cell is assumed to be 10, 15 and 20. A 4-path Rayleigh fading channel with uncorrelated scattering and *equal* average path power is assumed.

For the TCP ACKs an ideal scenario is assumed where the fixed sending hosts receive successfully the ACK packets generated by the mobile stations corresponding to each successfully transmitted TCP packet (i.e., no ACK packet is lost in the air interface and the wired part of the network) and that the ACK packets undergo no extra queuing delay in addition to the propagation delay. Packets corresponding to all the TCP connections are assumed to experience the same internet[1] delay ($D_{internet}$). An ideal fixed network behavior is assumed where there are no delay jitter and no packet loss for the fixed network part.

The values of some of the parameters used for obtaining the simulation results presented in this article are listed in Table 1. For detailed physical layer modelling and parameters please see [16]. With $M = 127$ bits, assuming that BCH codes are used, for $t = 1, 2, 3$ and 4, the number of information bits transmitted per frame-time would be 120, 113, 106 and 99, respectively. Therefore, for MSS = 1000 bytes, with (127,120,1),

---

[1]The term *internet* is used in a generic sense here.

(127,113,2), (127,106,3) and (127,99,4) BCH codes used as the radio frames, the number of radio frames corresponding to one TCP segment would be 67, 71, 75 and 81, respectively.[2] For an ACK packet size 52 bytes (40 + timestamp) and reverse link transmission rate of 16 *Kbps*, the ACK transmission time would be 26 *ms*. With $D_{internet} = 100$ *ms* and frame-time of 10 *ms*, the latency of transmission of a TCP packet would be therefore of the order of several hundred *ms*. The *ping* latencies in a 3G1X system were observed to be of the same order [13].

3.2 Results and Discussions

3.2.1 Radio Link Level Reliability and TCP Throughput

The radio link/PHY level target SIR significantly impacts the TCP throughput performance for both the rate adaptation schemes. Due to the TCP's inherent end-to-end error recovery mechanism, too much 'conservativeness' in terms of the target SIR, $(SIR)_0$ and hence radio link level reliability may not be desirable from the perspective of end-to-end throughput (Fig. 1a). Although the bit error rate (and hence the frame error rate) decreases with increasing $(SIR)_0$, the average per-connection TCP throughput (λ) may drop significantly at higher values of $(SIR)_0$ due to reduced radio link level throughput (as per [3], the upper bound of the rate capacity resource in a cell becomes increasingly tightened with increasing $(SIR)_0$). That is, at higher values of $(SIR)_0$ the impact of reduced frame error rate is masked by the smaller number of successfully transmitted radio frames, as a consequence of which, the number of TCP timeouts increases. This results in a reduced transmission rate at the sending host and hence reduced end-to-end throughput.

[2]For example, (1000 * 8)/120 = 67.

Table 1: Simulation parameters

Parameter	Value
Radio frame-length	10 ms
Radio frame-size, M	127 $bits$
Maximum spreading gain	128
Path-loss exponent	4.0
Variance of shadow fading	8 dB
Maximum normalized transmission rate, φ	8
Number of correctable single bit errors/radio frame, t	1, 2
Radio link level persistence parameter, R_{max}	4
TCP fast retransmit threshold	3
TCP segment size, MSS	1000 $bytes$
TCP ACK packet size	52 $bytes$
RTO_{min}, RTO_{max}, $RTO_{initial}$	1 s, 60 s, 2.5 s
Maximum TCP window size, W_{max}	8
Go-trip Internet delay, $D_{internet}$	100 ms

Again, for relatively smaller values of $(SIR)_o$ (e.g., 3 dB), although the radio frame transmission rate increases, increased frame error rate causes more radio link level frame retransmissions resulting in reduced radio link level (and hence transport level) throughput. Therefore, a concave behavior of the TCP throughput curve as a function of the target SIR is observed which results due to the competing behavior of the diminishing number of radio link level frame errors and the rate capacity resource in a cell.

The degradation in TCP throughput performance at higher values of $(SIR)_o$ is primarily caused due to the increased number of TCP timeouts (resulting from reduced radio link level throughput). As $(SIR)_o$ increases, the number of duplicate acknowledgments decreases while the number of TCP timeouts does not change much compared to the cases with smaller $(SIR)_o$. Reduced number of duplicate acknowledgments along with relatively higher number of TCP timeouts implies that the end-to-end error recovery is mostly triggered by the TCP timeouts and the TCP sender stays in the slow-start phase most of the time.

Since TCP throughput variation is not monotonic with $(SIR)_o$, an optimum desired $(SIR)_o$ can be found out which maximizes the average per-connection TCP throughput.

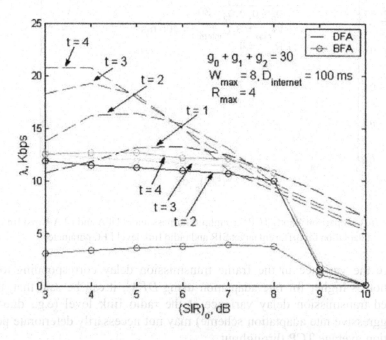

Figure 1a. TCP throughput comparison for radio link adaptation based on DFA and BFA for different target SIR and FEC coding (g_i denotes the number of TCP connections in cell *i*).

Under similar network conditions, TCP throughput performance is observed to be generally better with *DFA*-based dynamic rate adaptation. In particular, when the network load and/or the radio link/PHY level target SIR is high, *DFA* offers remarkably higher end-to-end throughput compared to that due to *BFA*. Also, with *BFA*-based rate adaptation, the end-to-end throughput decreases significantly when the value of the radio link level FEC parameter is small. Therefore, channel-gain based 'aggressive' dynamic rate adaptation would be desirable over less reliable radio links when the number of concurrent TCP connections is relatively large.

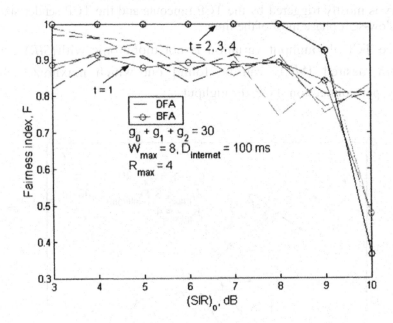

Figure 1b. Comparison among TCP throughput fairness under DFA and BFA-based link adaptation for different target SIR and radio link level FEC parameter

Since the variance in the frame transmission delay corresponding to a connection is higher for rate adaptation using DFA^3, it can be said that, the increased transmission delay variance at the radio link level (e.g., due to more aggressive rate adaptation scheme) may not necessarily deteriorate per-connection average TCP throughput.

3.2.2 Radio Link Adaptation and TCP Throughput Fairness

The TCP throughput fairness is observed to be better with *BFA*-based rate adaptation compared to that for *DFA*-based rate adaptation (Fig. 1b) except when $(SIR)_o$ is high and/or the FEC parameter t is small. In the latter cases, the average per-connection TCP throughput deteriorates significantly for rate adaptation using *BFA*. The reduced throughput fairness with *DFA*-based rate adaptation in the former cases is due to the more aggressive rate allocation among TCP connections with better channel conditions, as a consequence of which, the number of dropped TCP packets over a time

---

[3]This is due to the increased variance among rates allocated to a connection during successive frame-times.

period and the achieved TCP throughput becomes unevenly distributed over all the TCP connections. However, the fairness index with *DFA* is observed to be around or above 0.8 over the entire range of values of $(SIR)_o$ considered in this article. The throughput fairness generally deteriorates with increasing traffic load.

4. OUTLINE OF A GENERAL MODEL TO DESCRIBE THE INTERACTION BETWEEN TCP AND RADIO LINK CONTROL PROTOCOL

4.1 The Link Level Transmission Delay

For a persistent radio link control protocol, TCP timeout as a result of channel quality degradation and the consequent end-to-end retransmissions cause degradation in the end-to-end throughput performance. To describe completely the interaction of TCP with any infinitely persistent link layer error control protocol, we define the parameter *link layer transmission delay* ($tx_{LL}(n)$) which is the time to successfully transmit the nth TCP packet over the error prone wireless channel. Through the metric tx_{LL} the impact of a number of radio link level parameters can be lumped into one parameter whose interaction with TCP reflects exactly the link layers effect on TCP.

Under the assumption that there is no packet loss in the wired part of the network, we can calculate the throughput experienced during the nth interval as *throughput(n) = (TCP packet size)/tx_{LL}*. The average network throughput under steady state can be calculated using $E(tx_{LL})$.

Note that, when the wireless channel emulates a reliable link due to any infinitely persistent link level protocol, duplicate acknowledgments never occur. This is because packets are always delivered over the wireless channel in a FIFO fashion at the expense of large delays and round-trip time fluctuations. As no packets are delivered to the TCP receiver out of order, it never generates any duplicate acknowledgments. This then leads us to focus primarily on the TCP timeout mechanism.

4.2 TCP Expiration Value

TCP maintains a retransmission timer for every new sending window. That is, every time a valid acknowledgment is received, the timer is reset using an updated estimate of the round-trip time. If, however, the acknowledgment is lost, or the TCP packet that would have produced the acknowledgment is lost, a TCP timeout occurs.

We define the nth TCP expiration value $t_{TCP}(n)$, as the difference between the time the nth TCP timeout would occur and the time of arrival at the SR-ARQ module of the nth TCP packet minus any delays encountered in the network ($D_{internet}$) not including the wireless channel (i.e., $t_{TCP}(n) = t_{TCP}^{(timeout)}(n) - t_{TCP}^{(timeofarrival)}(n) - D_{internet}$). The TCP expiration value $t_{TCP}(n)$ depends on the mechanism that TCP uses to calculate its timeout value, which is based on the round-trip time corresponding to the previously transmitted packets. Essentially, $t_{TCP}(n)$ is the upper bound for the link level transmission delay tx_{LL} if we are to prevent a TCP timeout from occurring. Note that, both $tx_{LL}(n)$ and $t_{TCP}(n)$ constantly reflect the status of the network.

If we assume a constant value for $D_{internet}$, the round-trip time would depend only on past values of tx_{LL}, and therefore, $t_{TCP}(n)$ depends only on past values of tx_{LL}. This implies that if tx_{LL} increases gradually, $t_{TCP}(n)$ will also increase gradually, and it is more likely that $tx_{LL} \leq t_{TCP}(n)$ so that a timeout will not occur. If, however, tx_{LL} increases rapidly, there is greater chance that a timeout will occur.

It is also interesting to note that a timeout can be prevented by artificially manipulating tx_{LL} by gradually increasing it in anticipation of an actual steep increase in tx_{LL}. This allows the TCP expiration time to adjust and thus preventing a timeout.

4.3   TCP Expiration Envelope

We create the TCP expiration envelope (for *TCP Reno*) by plotting pairs of the TCP expiration value $t_{TCP}(n)$ and the link level transmission delay $tx_{LL}(n)$ (obtained by using *ns-2* [17]) against the time at which the start of transmission at the link layer occurs. In this case, TCP segment size = 1000 *bytes,* wireless link speed = 4 *Mbps,* buffer size at the BS = 50 *packets,* frame size = 256 *bytes,* and we assume that a selective-repeat ARQ scheme with a window size of 4 is used for radio link level error control.

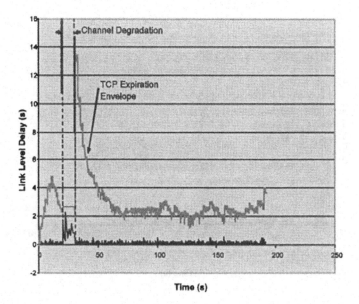

Figure 2. Link level transmission delay and TCP expiration envelope
(for FER = 0.05).

We artificially create sharp changes in channel capacity by initially running a simulation at a frame error rate (FER) of 0.01, and then between $20s-30s$, we increase the FER to some large value, and then back to the initial FER. Figs. 4 and 5 describe the situation where the inflated FER is 0.70 and 0.80, respectively. Note that, a timeout occurs in the latter case. The values of tx_{LL} remain at an increased value during channel degradation and then return to the values experienced before channel degradation. The number of TCP packets at the BS buffer increases during the periods of channel degradation.

Fig. 6 shows typical variations in the queue length at the BS with time, for the same situation when the FER is initially 0.01, and then it is inflated to 0.70 during 20s-30s, and then it returns to the initial frame error rate. Since an increased tx_{LL} implies a lower throughput, we expect to see a transient increase in the buffer size at the BS during the period of channel degradation. Any time afterwards the channel degradation period the buffer size remains more or less close to the original value.

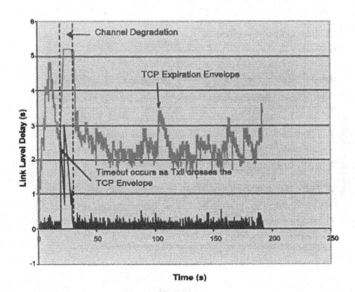

Figure 3. TCP expiration envelope for a channel fade between 20s and 30s.
(for FER = 0.70).

The sharp increase in the queue length is due to the TCP slow-start mechanism after a timeout during which the congestion window grows exponentially until it reaches the slow-start threshold *ssthresh* (which is 20 in this case). After the congestion window reaches *ssthresh,* the sender releases no more packets unless an acknowledgment is received, and in that case only releases one packet. In essence the number of packets outstanding is constant and is equal to *ssthresh.*

Since the rate of arrivals of acknowledgments varies and depends on the link layer transmission delay, it can be said that for a TCP flow the progression of data through the error prone channel determines the progression of data through the entire network. In the absence of any congestion loss, the throughput over the entire network is therefore equal to the throughput over the error prone channel.

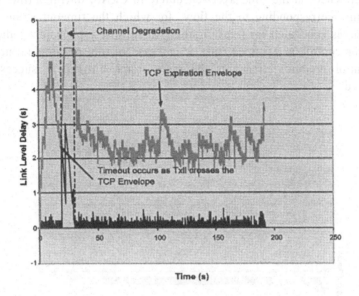

Figure 4. TCP expiration envelope for a channel fade between 20s and 30s (for FER = 0.80).

TCP timeout as a result of channel degradation and the consequent retransmissions cause degradation in the end-to-end throughput performance. This degradation can be alleviated by reducing tx_{LL} and using efficient scheduler based on relative values of the expiration envelope for the different flows.

5. CLAS: A CROSS-LAYER SCHEDULER FOR IMPROVING TCP PERFORMANCE IN WIDE-AREA CELLULAR 3G NETWORKS

Estimation of the TCP expiration envelope will enable us to design cross-layer optimized schedulers at the radio link level to minimize the number of TCP timeouts at the sender for the different flows. In this section, we propose such a scheduler, namely, CLAS (Cross-Layer Scheduler) which aims at minimizing the number of TCP timeouts at the sender, and thereby, increasing the TCP throughput (and hence wireless link utilization). Note that, in case of multiple concurrent TCP connections, the link layer transmission delay tx_{LL} is affected by the radio link layer scheduling policy.

We assume that the timeout value for each flow is available (or at least can be estimated) at the link layer scheduler. In CLAS, during a time slot, a radio frame corresponding to the flow, for which the timeout value is the minimum, is scheduled for transmission. An infinitely-persistent radio link layer error control is assumed, that is, in case of transmission failure, the error control module retransmits the radio frame until it is successfully transmitted.

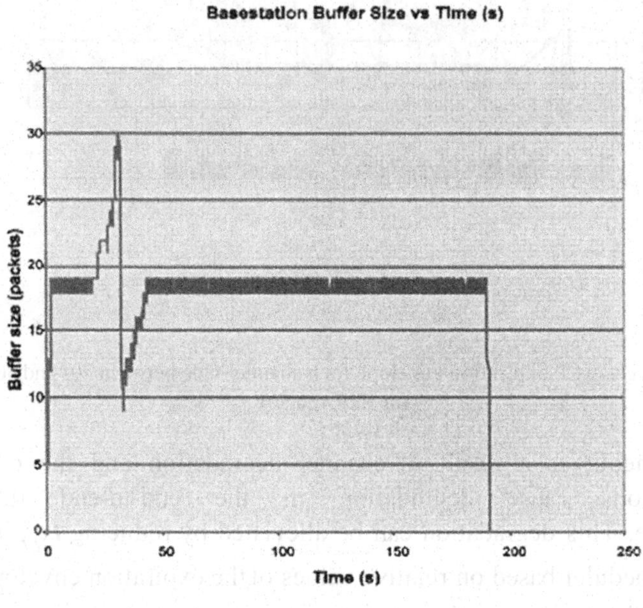

Basestation Buffer Size vs Time (s)

Figure 5. Variations in queue length at the BS when a timeout occurs

By using simulation, we observe the performance of CLAS compared to other scheduling schemes such FIFO and RR (Round-Robin) scheduling. In case of FIFO scheduling, a timestamp value (corresponding to the time of arrival of the earliest packet for a flow) is assumed to be available at the link layer scheduler. During a time slot, a radio frame corresponding to a flow for which the timestamp value is the minimum is scheduled for transmission. For RR scheduling, radio frames in the link layer queues (as described in section 2.2.3) are transmitted in a round-robin fashion. For all the three scheduling schemes, a radio frame corresponding to the flow, for which the link layer buffer is non-empty, is selected for transmission.

Preliminary performance results show that, with CLAS, for a relatively large number of concurrent TCP flows, the wireless link utilization (which is defined as the ratio of total TCP throughput for all flows and the link bandwidth) improves significantly compared to that for each of the FIFO and RR scheduling schemes (Figs. 7 and 8). The improvement in performance is observed to be increasingly better as the number of flows increases. We also observe that, with RR scheduling better channel utilization is achieved compared to that for FIFO scheduling. Note that, with CLAS, although the channel utilization improves, the TCP fairness performance may deteriorate compared to RR scheduling.

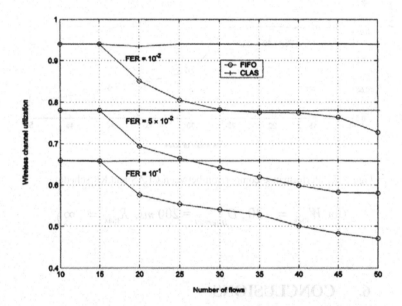

Figure 6. Performance comparison between CLAS and FIFO scheduler

(for $W_{max} = 20$, $D_{internet} = 200$ ms, $R_{max} = \infty$).

Note that, the relative performance of the link layer scheduling schemes are greatly influenced by the system parameters such as the maximum TCP window size (W_{max}) and the internet delay $(D_{internet})$. A detailed analytical investigation would be necessary to explore the inter-relationship among the TCP parameters and the scheduling parameters.

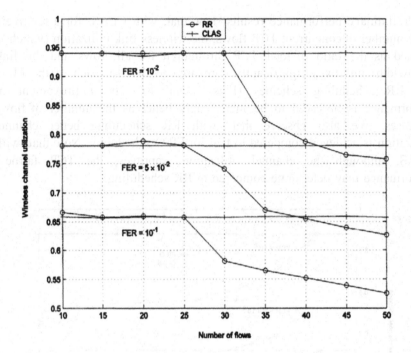

Figure 7. Performance comparison between CLAS and RR scheduler

$$(\text{for } W_{max} = 20, D_{internet} = 200 \text{ ms}, R_{max} = \infty).$$

6.   CONCLUSIONS

In this article, a multilayer model has been presented to investigate the impacts of the different radio link/PHY level parameters and adaptive radio link layer mechanisms on the end-to-end throughput and fairness performances in cellular WCDMA/3G networks. Based on this model, performance of TCP has been evaluated under dynamic radio link adaptation achieved through joint rate and power allocation for downlink data transmission in cellular multi-rate WCDMA/3G networks. To this end, we present the outline of a general model to describe the interactions between TCP and the radio link control protocol. The explanation of throughput across the network being equal to the throughput across the error prone channel along with our understanding of tx_{LL} in this model will allow us to describe completely the interaction of TCP with any infinitely persistent link layer error control protocol.

We have shown an example of how the transport layer information can be exploited to design radio link layer schedulers to improve network performance. A detailed analytical investigation on such a cross-layer scheduler is currently under progress.

REFERENCES

1. D. I. Kim, E. Hossain, and V. K. Bhargava, "Dynamic rate adaptation and integrated rate and error control in cellular WCDMA networks," *IEEE Transactions on Wireless Communications,* vol. 3, no. 1, Jan. 2004.
2. H. Chaskar, T. V. Lakshman, and U. Madhow, "TCP over wireless with link level error control: Analysis and design methodology," *IEEE/ACM Trans. Networking,* vol. 7, no. 5, pp. 605-615, Oct. 1999.
3. D. I. Kim, E. Hossain, and V. K. Bhargava, "Downlink joint power and rate allocation in cellular multi-rate WCDMA systems," *IEEE Transactions on Wireless Communications,* vol. 2, no. 1, pp. 69-80, Jan. 2003.
4. E. Ayanoglu, S. Paul, T. LaPorta, K. Sabnani, and R. Gitlin, "Airmail: A link-layer protocol for wireless networks," *Wireless Networks,* 1(1):47-60, 1995.
5. C. Parsa, J. J. Garcia-Luna-Aceves, "TULIP: A link-level protocol for improving TCP over wireless links," in *Proc. IEEE WCNC '99,* New Orleans, Louisiana, Sept. 21-24, 1999.
6. A. Chockalingam, M. Zorzi, and V. Tralli, "Wireless TCP performance with link layer FEC/ARQ," in *Proc. IEEE ICC '99,* pp. 1212-1216.
7. J. W. K. Wong and V. C. M. Leung, "Improving end-to-end performance of TCP using link-layer retransmissions over mobile internetworks," in *Proc. IEEE ICC '99,* pp. 324-328.
8. C. –F. Chiasserini and M. Meo, "Improving TCP over wireless through adaptive link layer setting," in *Proc. IEEE GLOBECOM, Symposium on Internet Performance (IPS 2001),* San Antonio, TX, Nov. 2001
9. C. –F. Chiasserini, M. Garetto, and M. Meo, "Improving TCP over wireless by selectively protecting packet transmissions," in *Proc. IEEE Conference on Mobile and Wireless Communications Networks (MWCN),* Stockholm, Sweden, 15 Sept. 2002
10. A. DeSimone, M. C. Chuah, and O. C. Yue, "Throughput performance of transport layer protocols over wireless LANS," in *Proc. IEEE GLOBECOM '93,* pp. 542-549, Nov. 1993.
11. Y. Bai, A. T. Ogielski, and G. Wu, "Interactions of TCP and radio link ARQ protocol," in *Proc. IEEE VTC '99,* pp. 1710-1714.
12. A. Chockalingam and G. Bao, "Performance of TCP/RLP protocol stack on correlated fading DS-CDMA wireless links," *IEEE Transactions on Vehicular Technology,* vol. 49, no. 1, Jan. 2000, pp. 28-33.
13. M. C. Chan and R. Ramjee, "TCP/IP performance over 3G wireless links with rate and delay variation," in *Proc. ACM MobiCom '02,* Sept. 23-26, Atlanta, Georgia, USA.
14. M. Zorzi, M. Rossi, and G. Mazzini, "Throughput and energy performance of TCP on a wideband CDMA air interface," *Wireless Communications and Mobile Computing,* vol. 2, no. 1, Feb. 2002, pp. 71-84.
15. *Computing TCP's retransmission timer,* RFC 2988, Proposed Standard, Nov. 2000.

16. E. Hossain, D. I. Kim, and V. K. Bhargava, "Analysis of TCP performance under joint rate and power adaptation in cellular WCDMA networks," *IEEE Transactions on Wireless Communications,* vol. 3, no. 3, May 2004.
17. S. McCanne and S. Floyd, "NS (Network Simulator)," 1995. URL http:/www.isi.edu/nsnam/ns.

Chapter 2

ALWAYS ON SERVICE INTELLIGENT NETWORK

SARIT MUKHERJEE, SANJOY PAUL, KRISHAN SABNANI
Bell Laboratories Research, 101 Crawfords Corner Road, Holmdel, NJ 07733, USA

Abstract: With the popularity of services like Push-to-Talk, the need for "always on" services is becoming important for service providers. This paper addresses the problem of supporting always on services in existing and new network architectures. It defines the requirements of always on services, identifies the problems in supporting such services, and proposes an overlay network based solution to make always on services real. Some results from initial prototyping and experimentation are also presented to demonstrate the feasibility of deploying such services.

1. INTRODUCTION

Overwhelming success of Instant Messaging (IM) made it clear that there is an inherent appeal in being able to communicate "instantly" at the click of a mouse. This was further confirmed by the huge success of Nextel's Push-to-Talk (PTT) feature which enabled Nextel subscribers to communicate using a nation-wide walkie-talkie at the push of a button. Note that PTT succeeded even when people could call anyone anytime from their cell-phones. There are several reasons why PTT is preferred over a phone call:

- PTT connection time is less than a second while the connect time for a cell-phone call can be as much as 5 seconds, if not more.
- PTT does not require the user to do anything more than pushing a button as opposed to dialing a phone number.
- PTT uses Voice over IP (VoIP) instead of circuit-switched voice and hence is cheaper and feature rich.

The critical observation is that both in IM and in PTT, the user is "always on" [1] in the sense that the user is always connected to the network; network knows the presence of the user and hence keeps the user's state and the user knows it is connected to the network and hence does not need to explicitly dial the network to set up a call.

While the term "always on" has become popular, there is no formal definition of "always on" in the technical community. We define a service to be always on if:

User:

  (i)   does not feel disconnected anytime, anywhere
  (ii)   can access the service instantly regardless of location and time
  (iii)   once connected receives desired quality of service and improved experience
  (iv)   is always in control; is always protected and is always engaged [2].

Network:

  (i)   can reach an end-user instantly at any time regardless of the user's location
  (ii)   can push info to end-user regardless of the state of the mobile (active/ dormant)

1.1 Motivation and Problem

While the concept of being always on seems inherently appealing, it becomes increasingly difficult to provide such a perception to a user in a mobile wireless network because of the following reasons:

  (i)   Choppy network connection: Wireless links are inherently error-prone because of multi-path and fading. A noisy link leads to connection drop making always on a difficult proposition.
  (ii)   Inadequate coverage: All areas are not covered equally well by a wireless carrier. Therefore, a mobile user can get disconnected once it moves from an area of better to a poorer coverage.
  (iii)   Mobility in dormant state: Locating the user becomes difficult especially when the mobile is dormant and as a result when a dormant mobile becomes active far from where it went dormant, the network has to spend some time locating the user and that makes "instant" connection almost impossible.

The challenges can then be summarized as:

(i)   Reduction of connection time: The network must maintain user's state and should not tear it down even when the user is dormant. In addition, the network must keep track of the user regardless of whether the mobile is active or dormant and keep transferring its state to the closest network element, if, need be, so that connections are never broken.

(ii)   Improving the quality: It is not enough just to have "continuous" connectivity because a poor-quality connection would leave the user dissatisfied. Thus the goal in always on service is to provide and maintain "improved" service quality. Moreover, the same quality should be maintained for the user even when the mobile comes out of dormancy.

(iii) Richer functionality: Since the network knows the user's location and presence it would be possible to provide an "enhanced" experience to the user or to provide the user some service that is not otherwise possible.

(iv)   Pushing information: In contrast to existing mechanisms in which the user initiates a connection to the network, the network has to initiate a connection to the user. This requires additional "smarts" within the network.

2. SHORTCOMINGS OF EXISTING NETWORK ARCHITECTURE

Existing networks are not built to support always on services. Figure 1 shows the typical steps a client, network and a server go through before executing any transaction in a legacy network (without always on) and in an always-on network. For example, in a legacy network, the client goes through a set of device-related steps such as powering up the device, booting up the operating system; network-related steps such as setting up connection (PPP connection), authenticating the user; and application-related steps such as starting the application (such as Outlook or Internet Explorer), connecting to the server (Outlook server or eBay server for example), authenticating with the server and accessing the service. The network also goes through air-interface related steps such as selecting and acquiring the channel; device-related steps such as authenticating the device (using IMSI or device-specific characteristics) and registering the device; network-related steps such as setting up link-level connection (PPP connection), and allocating IP addresses. The server authenticates the user, sets up connection and responds to client request. All these steps in a legacy network add up to several minutes and as a result, the user feels "disconnected" and the process of doing a transaction becomes very cumbersome. Always on network simply eliminates many of these time-consuming steps (shown by striking out itemized steps in Figure 1) resulting in a minimal number of steps (shown at the lower part of Figure 1) that are needed for connecting a client to a service. As a consequence, the process of connecting to a service is simplified and the time is reduced to a few seconds or even less than a second in some cases and hence the user does not feel disconnected.

As shown in Figure 1, to make a service always on, each one of the client, network and the server has to do something to eliminate delays. This paper focuses on the network aspects.

There are several other drawbacks of the existing networks:

(i)   Inadequate wireless coverage leads to disconnection: When a mobile enters a tunnel or a building where the wireless network signal is not strong, the data connection drops. The user has to go through the entire connection establishment phase once he gets out of the low-coverage area.

(ii)   Connection state maintenance: In order to provide always on connection, the network has to maintain state of the mobile's connections even when the mobile is not active. This implies that a carrier's network has to potentially maintain states of as many mobiles as there are subscribers. This number can potentially go into tens of millions and the current network elements are not equipped to handle that.

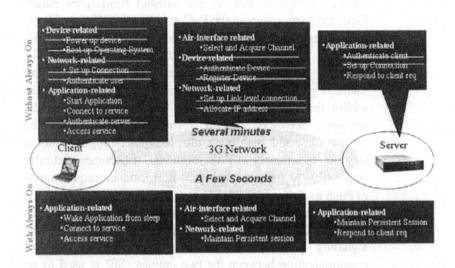

Figure 1 Always On eliminates time-consuming operations of a Legacy Network

(iii) Making applications "always on" one at a time: In order to enable always on service, the carriers are putting intelligence in the servers. For example, Qualcomm's QChat server which is used to provide PTT service maintains user's state and location information. Instead of incorporating the same features in each server independently, it makes more sense to incorporate them once and for all in a gateway that every server (potentially belonging to different services) can leverage.

(iv) Making applications "always on" one Network at a time: While carriers are making applications always on, the scope of always on is limited to a single carrier's network and also it is limited to one "type" of network. For example, Nextel's PTT service works only within Nextel's network and it does not work when the mobile moves into an area covered by WiFi hotspots but not covered by Nextel's network. Ideally, the always on service should be provided across different types of networks (3G, WiFi, 4G) and across different carriers' networks. Open Mobile Alliance (OMA) consortium is pushing Push to Talk over Cellular (PoC) [10] which is an open standard using SIP signaling protocol aimed at making Push to Talk interoperable across carriers. However, PoC is not without limitations either. First, the participants in the PoC effort have different views. For example, Nextel wants proprietary technology within a carrier's network and wants to standardize across the carriers' networks at the edge. Ericsson, Nortel and Siemens are pushing for a completely open solution: open within the carrier's network as well as open across the carriers' networks. As a result of opening up the architecture, and enabling multiple types of applications to share a common set of IMS resources, it becomes virtually impossible to meet the stringent performance requirements of Push to Talk service in the PoC architecture.

(v)   IMS infra-structure has Go/Gq interface between PDSN/GGSN (on the bearer path) and PDF (on the signaling path) but there is no standardized way of communication between the two entities: SIP is used to set up connections and once the connection is established, mobiles exchange data using the bearer path. However, if the data connection breaks, there is no feedback between the bearer path and the signaling path and as a result, the connection remains on by default until an administrator removes those connections via a manual operation. Ideally, there should be an explicit feedback between the bearer plane and the signaling plane such that the network element on the bearer plane, on behalf of the mobile, can keep the SIP connection on even when the data connection is broken because of poorer coverage, for example.

(vi)   Support for Push: In the current network, a client has to initiate a PPP connection with the network and once the connection is established, the client is able to pull content from the network or from servers outside the network. However, if there is no explicit mechanisms for the network to initiate a connection with the client and push content. Recently, 3GPP working group has proposed mechanisms such as Network Requested PDP Context Activation (NRPCA) [7] and Multimedia Broadcast / Multicast Service (MBMS) [8, 9] to enable network-initiated push functions.

3. PROPOSED SOLUTION

There are various ways in which an "always on" service can be provided in a carrier's network. However, there are two important points to keep in mind:

1. If the goal is to provide always on service across multiple networks, then we need a "network" independent architecture.
2. If the goal is to provide always on service for a particular carrier's network, then various network elements in the traditional carrier's network need to be augmented and that may delay deployment of such a service.

In order to facilitate (1) and avoid the drawbacks of (2), we propose to use an "overlay" network which is independent of the underlying network technology (3G, 4G, WiFi) and the administrative boundaries (Verizon Wireless, Sprint PCS, Cingular, etc.). Naturally, the overlay network can provide always on service either to the customers of a specific wireless carrier, such as, Verizon Wireless across disparate link layer technologies like 3G and WiFi networks; or to the customers of a Mobile Virtual Network operator (MVNO) across multiple 3G wireless carriers' networks.

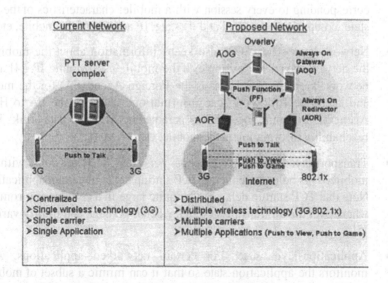

Figure 2 Components of an Always On Network

The overlay network is created using two network elements, namely, Always on Gateway (AOG) and Always on Redirector (AOR). The goal is to bundle "all the always on specific functionalities" into the AOG so that the existing network elements in a carrier's network do not need to be changed. In addition, the purpose of AOR is to haul traffic from various access networks into the overlay network transparent to the end user. Once the traffic is hauled into the overlay network, AOG can provide all the necessary always on functionalities. In addition to that, a Push Function (PF) is introduced in the architecture to support network-initiated "push" functions on the overlay network. A typical always on overlay network is shown in Figure 2. Here is a description of the functions provided by the network elements in the overlay.

Always on Gateway (AOG): The AOG is the core of the overlay. One or more AOGs are used to build the overlay. Each AOG performs the following functions.

1. Maintain state information: AOG keeps multi-layer state information of a large number of mobiles that it services.

   - Data link-layer state: The mobile establishes a point-to-point protocol (PPP) [3] session with the AOG. PPP sessions are state-full in the sense that the AOG needs to keep the following session information corresponding to every session with a mobile: characteristics of the link state, compression protocol and its state, IP address of the mobile, etc.

   - Network-layer state: AOG also keeps information about the mobile at the network layer. For example, if the mobile uses Mobile IP [4] at the network layer, the AOG acts as the Foreign Agent (FA) for the mobile and keeps the following state information: states for the FA to Home Agent (HA) tunneling, IPSec parameters, if any, data link layer reachability information to the mobile, etc.

   - Transport-layer state: AOG keeps persistent TCP sessions with the mobile. This helps in reducing TCP startup delay in certain applications. Note that TCP startup delay can be quite large in a cellular environment where over the air transmission delay could be large and highly variable [5].

   - Application-layer state: For certain networked applications, AOG monitors the application state so that it can mimic a subset of mobile's behavior to a server in case the mobile becomes temporarily unreachable due to unavoidable network conditions (for example, no coverage within a tunnel).

2. Migration of state information: A mobile's state is initially created in an AOG when the device is powered up. As the mobile roams in the underlying network, and goes far away from the AOG that has its state information and comes closer to another AOG, then for better service, the whole state information can be transferred from the previous AOG to the new AOG. This eliminates the unnecessary session setup with the new AOG but still keeps the mobile always on.

3. One time authentication: In today's network, when a mobile roams from one network to a visiting network (where network boundary may be defined across different carriers or across different physical-/data-link-layers), usually the mobile and/or user must be authenticated by the visiting network, even if the mobile and/or user was authenticated by the previous network. In the overlay network architecture, AOG can authenticate the mobile and/or user once and make roaming across different networks seamless and fast.

4. Enhance services for the mobile: Since "over the air" communication is expensive compared to communication over the wired network, AOG cuts down on "over the air" communication by proxying on behalf of the client and performing the client functions in the wired network. This reduces the session setup time, eliminates or reduces round trip time to execute DNS queries, etc.

5. Multimodal operation: The overlay network can manage multiple overlapping networks. For example the underlying networks could be a WiFi hotspot and a wide area 3G network. In the hotspot region, both networks are accessible by a mobile equipped with multiple interfaces (e.g., built-in WiFi interface and 3G PCMCIA card). In such a scenario, the common wisdom is to have intelligence in the mobile to switch to the high bandwidth network (i.e., WiFi hotspot) and stick to it as long as the quality of reception is good. In the overlay network environment, we let AOG control which network interface should be used by the mobile. There are several advantages to this approach. First, since the intelligence is moved from the mobile to the network, the end device can be made simple. Second, the AOG can seamlessly use the high bandwidth network in the middle of an application session between the mobile and a server. Third, even if the hotspot may provide higher data rate, it could be temporarily overloaded. In such a situation the mobile would be better off using the wide area network. AOG can determine this very easily and efficiently and direct the mobile accordingly.

6. Context notification: AOG monitors roaming of a mobile within a network or across different networks, and dynamically prepares context information about the mobile. The context information includes parameters like network identification, available bandwidth, IP addresses of the intermediaries (e.g.,

local web object cache, local SIP proxy, local DNS server, etc). AOG sends the context information to the mobile and the servers that subscribe to such messaging. The mobile can use the context message to configure itself. For example, the mobile can reconfigure its web browser to explicitly use a proxy cache in the AOG. This improves web browsing experience by keeping persistent session with the local cache and reducing "over the air" DNS operations. Similarly the SIP proxy address can be configured into the mobile to reduce DNS access over the air. The server may use the information to provide personalized, location-dependent, bandwidth-sensitive services.

7. Overlay network setup: AOG sets up an overlay network with other AOGs. The details of overlay setup are omitted from this paper, but the interaction between AOGs is shown in an example application later in the paper.

8. Location and paging of mobiles: AOG keeps track of the mobile as it moves across multiple networks. Note that a mobile can either move across networks with the *same* physical-/data-link-layer (e.g., 3G) owned and operated by *different* carriers or move across networks with *different* physical-/data-link-layer (e.g., 3G and WiFi). AOG keeps up-to-date location information in both the above scenarios. In some networks, to reduce network load, paging may also be initiated by AOG to find a dormant mobile.

Always on Redirector (AOR): AOR is used to redirect traffic from the mobile to the "best" AOG, based on location of the mobile, QoS requirement of the session, home network of the mobile, etc. An AOR interfaces with a particular access network (e.g., 3G, 4G, WiFi, etc.). While there may be multiple AORs interfacing with a particular access network, there cannot be a single AOR interfacing with more than one access network.

An AOR has two interfaces, one facing the overlay network (AOG in particular), and the other facing the access network from which traffic needs to be hauled into the overlay network. The interface facing AOG implements signaling and bearer sessions following the 3GPP2 standards (A10-A11) [6]. In other words, a PPP connection originating from a mobile must terminate at AOG thereby giving AOG the full control of the connection including assignment of IP address. In order to achieve this in a uniform manner across different networks, the functionalities of the interface of the AOR facing the access network are customized based on the characteristics of the access network. We briefly describe this feature of AOR for two types of access networks: CDMA2000 network and a WiFi network.

In order to interface with a CDMA2000 network, AOR acts as a PDSN to the PCF of the CDMA2000 network, and acts as a PCF to the AOG. AOR takes a mobile's session from the PCF and redirects it to an AOG depending on the service attributes of the mobile. In other words, if the mobile has subscribed to the always on service, then the connection is redirected to an AOG, otherwise the connection is passed on to the PDSN. Therefore, it is possible to offer always on service only to subscribers that sign up for it, and leave aside the remaining subscribers. To achieve this, AOR proxies as a PDSN and terminates A11 signaling from PCF, and then terminates Link Control Protocol (LCP) phase of the PPP negotiation with mobile. Then AOR initiates user authentication with the mobile during which the user identification is sent. AOR uses the user identification to determine if always on service is to be provided to the user. If yes, the AOR opens a new session with the AOG; otherwise it opens a new session with the PDSN in the CDMA2000 network. In both cases, AOG or the PDSN completes the data call with the mobile and AOR splices the two PPP sessions: one between the mobile and the AOR and the other between the AOR and the AOG/PDSN.

When AOR interfaces with a WiFi network, it sits upstream of an Access Point (from the mobile's point of view) and implements a PPTP proxy (or L2TP proxy). All the mobiles in the network get configured with the IP address assigned by the Access Point. The mobiles that subscribe to the always on service initiate a PPTP session with AOR as the PPTP server. This forces the mobile to tunnel PPP frames to the AOR using the IP address provided by the access point as the source address. AOR, instead of terminating the PPP session, tunnels the PPP frames all the way to the AOG responsible for terminating the session and assigning IP addresses. In this case the mobile is assigned two IP addresses: the address assigned by the Access Point is used for tunneling PPP frames while the address assigned by the AOG is used for all data communications. Again, in this case, always on service is provided only to a mobile that subscribes to the service and the rest are left untouched.

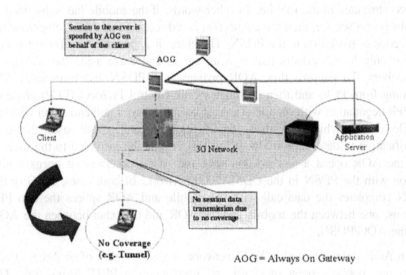

Figure 3 Architectural view of Network-initiated Smart Push

Push Function (PF): Push Function is used by the network to facilitate network-initiated content push to the user. PF interfaces with the Application Server on one side and with the AOG on the other side. Push Function is triggered by an Application server to initiate a push operation. However, PF determines how the content will be pushed. For example, content may be pushed using SIP in the IMS architecture [7] or using Multimedia Broadcast Multicast Service (MBMS) [8] or using WAP [7]. PF together with AOG can provide interesting capabilities, such as "smart" push to a wireless network. The idea behind "smart" push is to prevent a network from pushing content to a mobile end user who may not be temporarily reachable and hold the content in the network until either the mobile becomes reachable when the content is delivered or the content loses relevance. Architectural view of "smart" push is shown in Figure 3.

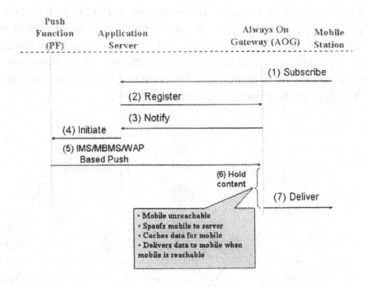

Figure 4   Call Flow for Network-initiated Smart Push

Call flow corresponding to smart push is shown in Figure 4. First the mobile subscribes to a service, such as "location-based smart push". Application server registers with AOG (Step 2) by providing the conditions under which the AOG should send it a notification. For example, in a location-based service, the application server could be a "coupon" pusher for Macys and the condition could be the mobile being in the vicinity of a Macys store. When the mobile is in the vicinity of a Macys store, AOG notifies (Step 3) the application server. The application server sends a message to PF to initiate the push. PF decides the best technique for pushing the content based on the content itself, the capability of the mobile station and the available network bandwidth (Step 5). AOG intercepts the content and holds it (Step 6) if the mobile is not reachable due to poor coverage while spoofing on behalf of the mobile client to the Application server. When the mobile becomes reachable (before the content loses relevance), AOG delivers the content to the mobile (Step 7). If the mobile is unreachable for a long duration such that the content loses relevance, AOG simply drops the content thereby saving network bandwidth.

3.1   Example: Always On Multi-party Chat

Using always on multi-party multi-network chat as an example application, we show the high level call flow for the overlay network. The overlay network is built on top of multiple access networks as shown in Figure 5. We show both wireline and wireless networks in the access to show that AOR can indeed interface with networks with different physical-/data-link layer technologies. The application is a multi-party chat where users join the session on their own or are invited to the session by a participant. The session establishment uses SIP-like protocol. The steps in the example are described below:

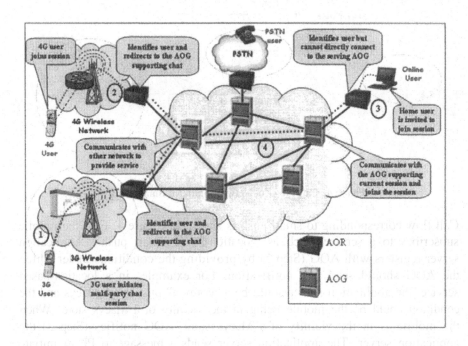

Figure 5 Call Flow in an Always On Network supporting Multi-party Chat

1. A 3G user initiates a chat. The AOR associated with the 3G access network recognizes this and redirects the call to the nearest AOG supporting the multi-party chat application.

2. A 4G user joins the session. The AOR associated with the 4G access network recognizes this and connects the user to the nearest AOG supporting the multi-party chat application. In this example, the AOG selected by the AORs happens to be the same.

3. A user at home using a cable modem is invited to join the session. The AOR associated with the cable modem access network recognizes this and redirects the user's connection to the nearest AOG. In this case the AOG selected is not the same as the one above.

4. Now the selected AOGs communicate with one another on the overlay network to bring all chat participants to the same chat session.

Note that the components of the overlay network participate both in signaling and bearer path (unlike SIP architecture where they are separate). The advantages of having them in both paths result in better management of state information and bearer path quality provisioning.

3.2   Prototype Implementation

In this section we describe results from two prototypical implementations and experimentations. First one shows how keeping persistent session with AOG reduces service time for web browsing giving the user always on experience. The second one describes implementation of AOR on Linux and shows how fast it redirects connection even though it functions in the overlay network.

3.2.1   Response Time Reduction with Session Persistence

We implemented prototypes of session level optimization techniques in Linux and conducted some controlled experiments to measure the quantitative benefits of session persistence. In this section we present the experimental setup and the summary of the results obtained from the experiments we conducted.

The experiments are conducted for mobile web browsing. The browser of the mobile can be set to point to an (explicit) proxy cache collocated with the AOG. A persistent TCP session is maintained between the browser and the AOG. To perform the experiments with specific web pages in a controlled environment, the top level pages and the embedded objects in them from the web sites were copied to a local apache web server. For this experiment, the top level pages from Yahoo, CNN and Britannica were copied. The statistics for these sites are:

- Yahoo (www.yahoo.com): It has 16 embedded objects hosted in 3 different domains. The size of the page is 74 KB. This constitutes a typical web site with small number of domains.

- CNN (www.cnn.com): It has 58 embedded objects hosted in 6 different domains. The size of the page is 197 KB. This constitutes a typical web site with medium number of domains.

- Britannica (www.britannica.com): It has 32 embedded objects hosted in 15 different domains. The size of the page is 178 KB. This constitutes a typical web site with large number of domains.

The browser at the mobile was instrumented to compute the time between the sending of the request for the top level page and the complete display of the page (including all embedded objects). We refer to this as the *user perceived* response time for the page. We measured this response time at the browser to download three popular top level pages (Yahoo, CNN and Britannica) and the embedded objects contained in them. The results are shown in Figure 6.

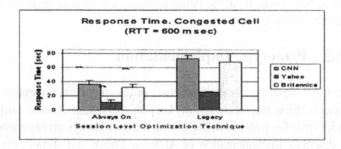

Figure 6 Improvement in Response Time with Always On

The response times were measured and averaged over 20 downloads of each of the top level pages and the corresponding embedded objects. The legacy case is where the mobile sends all the DNS requests over the air. Moreover, no persistent session is maintained which results in TCP startup delay for every TCP connection established by the mobile. In the always on case, the mobile maintains a persistent session with AOG and all the DNS resolutions are done by AOG on the wired network.

As is evident from Figure 6, eliminating over the air DNS queries significantly reduces the response time. In addition, response time is improved by eliminating the need for re-establishing TCP connections for every web session. Response times for CNN and Britannica are much higher than that for Yahoo because Yahoo has fewer domain names and embedded objects than the other two. Between CNN and Britannica, CNN has more embedded objects but fewer domains. The time required for making DNS requests balances out the time required to download the embedded objects and as a result, they have similar response times. In all cases, the response time for always on sessions is much smaller than that for legacy sessions. In fact, we observe a mean response time improvement of 50% for congested cells.

3.2.2 Price of Overlay

In order to estimate the overhead for using an overlay network, we built an AOR and measured the additional latency introduced due to overlay. AOR prototype for CDMA2000 access network was built on Linux 2.4.18 kernel running on a Pentium III 450MHz workstation. Different modules of the prototype are shown in Figure 7. The Ethernet Frame Processor and the IP Packet Processor are already provided in the standard Linux distribution. We implemented the rest of the modules as a patch to the kernel. Brief description of each of these modules is given below:

- UDP Processor: This module is responsible for sending and receiving UDP packets.

- R-P Signal Processor: This module is responsible for R-P session termination between PCF and AOR, R-P session establishment between AOR and AOG and for splicing them.

- AAA Client: This module implements a RADIUS client to access a AAA server for authenticating a user and to get the IP address of the destination AOG.

- GRE Processor: This module takes a PPP frame from the PPP Control Processor, encapsulates it and sends the frame in a GRE session. It also de-capsulates a PPP frame from a GRE packet and delivers the frame to PPP Control Processor. It is also responsible for splicing the PCF-AOR and AOR-AOG sessions using the GRE keys.

- PPP Control Processor: This module is responsible for terminating a mobile initiated link control phase of a PPP session, identifying the user from PAP/CHAP, authenticating the user, getting AOG IP address from a AAA server using the AAA client, and for initiating link control phase of a PPP session with the selected AOG.

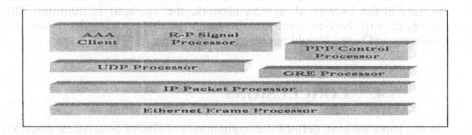

Figure 7 Prototype Implementation of AOR

We conducted a set of experiments to measure the additional overhead AOR introduces in setting up a data session between a mobile and the AOG. We measure the delay introduced in the data or bearer path (i.e., actual packet transfer).

Figure 8 shows the transfer latency of a data packet (averaged over multiple runs) with and without AOR. A packet size of 1300 bytes (i.e., the MRU of the PPP session) was used. In order to create the background load on AOR, a number of PPP sessions each sending a constant bit rate traffic to the AOG were generated. By changing the number of these sessions background load was varied. Packet transfer latency was computed by tagging the reception of every packet. Note that the packet transfer time increases with load, however the overhead introduced by AOR is negligibly small.

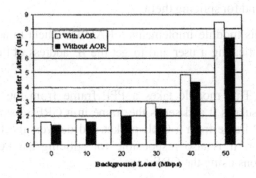

Figure 8   Packet Transfer Latency with and without AOR

In addition, Figure 8 shows that in the worst case, the connection setup time and the packet transfer delay with AOR is only a few milliseconds larger than in the case without AOR. This shows even though AOR redirects packets in the overlay always on network, the overhead is minimal. This overhead will become even smaller with product quality hardware and software and hence should not be a cause for concern.

4.   CONCLUSIONS

In this paper we have defined the requirements of always on services, described the major problems in supporting such services in today's network, proposed a network independent solution that provides the benefits of an always on service without changing existing networks, and have also presented some preliminary results from our prototype implementations to show the benefits of always on services. In addition to that, we have shown how novel services, such as "smart" push can be provided a service provider by combining the control-plane intelligence, such as user-profile and policies with data-plane intelligence, such as reachability of the mobile user. Currently we are optimizing the implementation of AOR and AOG for the basic function and are enhancing the functionality of these elements to provide an "enhanced" always on service experience to the end users.

REFERENCES

[1] 3rd Generation Partnership Project 2. Interoperability Specification (IOS) for cdma2000 Access Network Interfaces --- Part 3 Features. 3GPP2 A.S0013-A, Version 2.0.1, July 2003.

[2] America On Line. http://www.wave-report.com/other-html-files/alwayson2003.htm

[3] W. Simpson et. al. The Point to Point Protocol (PPP). Internet Engineering Task Force (IETF) Request for Comments (RFC) 1661, July 1994.

[4] C. Perkins et.al. IP Mobility Support for IPv4. Internet Engineering Task Force (IETF) Request for Comments (RFC) 3220, August 2002.

[5] P. Rodriguez, S. Mukherjee and S. Rangarajan. Session Level Techniques for Improving Web Browsing Performance on Wireless Links. In Proceedings of the Thirteen International World Wide Web Conference, New York, May 2004.

[6] 3rd Generation Partnership Project 2. Interoperability Specification (IOS) for CDMA2000 Access Network Interfaces --- Part 7 (A10 and A11 Interfaces). 3GPP2 A.S0017-0 v2.0, May 2002.

[7] 3rd Generation Partnership Project; Technical Specification Group Services and System Aspects Push Architecture (Release 6). 3GPP TR 23.976 v2.0.0 (2004-03).

[8] 3rd Generation Partnership Project; Technical Specification Group 22: TSG 22.146 "Requirements for MBMS".

[9] 3rd Generation Partnership Project; Technical Specification Group 23: TSG 23.246 "Architecture and Functional Description".

[10] Open Mobile Alliance (OMA). Push to Talk over Cellular Requirements; Draft Version 1.0 - January 31, 2004. OMA-RD_PoC-V1_0-20040131-D

Chapter 3

COGNITIVE TRENDS IN MAKING
Future of Networks

PETRI MÄHÖNEN
Aachen University, RWTH Aachen, Wireless Networks Group, Kackertstr. 9, D-52072 Aachen, Germany

Abstract: We are arguing that some new interesting trends in making are emerging. In this paper we try to identify some aspects that could provide hard problems with potentially high payout for the future research. We also show how some of the topics are shared between both wireless and fixed network domain. We propose as a hypothesis that the key issue in the future is to understand and lower the increasing complexity (increasing abstraction) of communications systems. There are several different ways towards this goal.

1. INTRODUCTION

For this contribution we have chosen few possible *trends-in-making* that could have a relatively large impact on the future research of wireless and fixed networks. This selection is focused to few possible long term issues, and as such we do not claim to provide comprehensive roadmap of the future technologies. The selection is in large part merely author's personal guesstimate on emerging and interesting research lines that have been suggested. One could also see this invited contribution as (in part) an extension to earlier work[1], which was more firmly based to heterogeneous networking. There are number of interesting reports and works describing the possible future R&D issues, among them we can refer to excellent treatments by Raymond Steele[2], David Farber[3] and to the NSF report of the future of networking research[4].

We focus mostly on wireless communications and networking perspective. The discussion is aimed to look at trends and architectural principles, so we are not rich on details – taking into account the speculation required to looking for future trends we think that this is somewhat acceptable for invited "future trends" paper. Our main premise is that different adaptive and intelligent approaches will become part of wireless and fixed network architectures. The expressed thinking and work has been influenced and derived from a large part from the viewpoint of wireless communications and software radio research. The seminal work, and highly recommendable paper, approaching the same idea space as ours through innovating *knowledge plane* from the Internet perspective is described by D. Clark *et al.* [5].

2.   COMPLEXITY AND TUSSLES

2.1   Status Quo

The cellular networks, unlicensed short range communications networks, and most notably Internet have been great successes. The incremental works in different subtopics are making these successes even greater and pervasive. However, as has been noted in several contributions[4,5], we should be careful on not to abandon more fundamental and disruptive research.

The success of wireless networks and Internet has been driven by many reasons. However, one can claim that in large part the relatively clean and simple architecture has been one of the key ingredients for the success. Both systems have also been providing a relatively easy access to end-users. Especially terminals of mobile communication are typical consumer equipments (mobile phones), i.e. very easy to use, although the technology inside is relatively complex. However, one should note the potential "VCR-clock syndrome"; as the number of features increases with mobile phones ever larger portion of users are not using those or even are not aware of all capabilities.

The number of mobile phones has already exceeded 1 billion worldwide, and also current IPv4 based Internet has become a huge and relatively complex network to comprehend (in part since the end-to-end principle is not anymore fully adhered due to e.g. NATs). In short, networks are becoming larger, there is tendency to glue them together (like all-IP approach), and stakeholders try-and-add ever increasing number of features into them (like QoS, support for multicasting, labeling etc.). All of these interesting and valuable issues are omitted in the following discussion, and we try to look possible larger and more futuristic trends, which are partially emerging due to present development and trends.

2.2 Complexity

The complexity of networks, especially at the global scale, is increasing rapidly. Although architectural principles are followed and adhered to with reasonable accuracy, inevitably systems have become complex, include some on-the-spot stop-gap solutions (even inelegant ones), and overall are becoming harder to understand. We have successfully used layering and abstraction to design and manage our present systems. There is a current trend emphasizing cross-layer optimization. This shows that there is a tendency to see benefits on violating this compartmentalization. The seriousness of 'violation' depends, however, very much on whether one is really violating layering on data plane or merely introducing some more clever use of control plane(s).

In fact, one can argue that if the visions on sensor networks with connectivity to other core networks become reality within 10 years combined with ever increasing number of other networked (wireless) devices[15], the overall network becomes as complex as some biological systems. Trillion devices networks might lay in the foreseeable future. When this is combined with the economical and trust-reputation-privacy related issues, we might experience complexity crises - at least on the level of fundamental research when we are trying to understand these systems. Due to sheer size and diversity of components in such ultra-large networks the IPv6 is not a complete answer for our future needs. It might become tempting to start to rethink our architectural principles and seek if we need completely new insight (we have occasionally called this X-Net in our internal projects).

One of the likely emerging trends in making will be to trying to understand complex networks at the fundamental level (where and how does the complexity emerge, how do these systems behave) and designing new abstractions to fight against critical complexity.

One of the first indicators for this trend has been the immense revitalization of network topology research, as started by the work on small worlds and scale-free networks. While especially the small world -concept addresses one facet of locality in networks (in terms of clustering coefficient and characteristic path length scaling), there is still room for more topology research to produce different ways of abstracting network topology information. We have to understand better the dynamics of network formation and growth, especially when wireless communications and various autoconfiguration techniques are being applied. Actually one trend in making might be the vigorous increase of the fundamental network research, including the fundamental understanding of network topology formation. I would dare to argue that as the number of independent, but interacting, nodes is increasing many classical physics methods are becoming valid for modeling at least partial behavior at large-scale. Some initial results along this line by different groups (Aachen among them) have been quite promising.

2.3 Tussle in Wireless and Cyberspace

Clark *et al.*[7] have described tussles in Cyberspace, i.e. the fact that different stakeholders have sometimes highly different interests and requirements for their needs. Their description is focused to fixed Internet, but it should not become as a surprise that same arguments are applicable to wireless networks. In fact, as the wireless systems are moving more towards providing data services and voice-only paradigm is declining, the emergence of tussles in wireless might actually become even stronger than in the fixed cyberspace. The large problems generated by 3G spectrum auctions and slow acceptance of 3G networks in Europe can be seen as symptoms of these underlying tussles.

One of the problems is that traffic patterns and key economical assumptions are not generally well understood and studied. A good example is the firmly held believe that "content is the king" without any strong proofs, as Andrew Odlyzko[8] has eloquently described. The highly interesting papers from Odlyzko[8,9] are showing that we should be more careful with our analysis and assumptions, especially when reading white papers. We are also arguing that one should be quite cautious before emphasizing *real-time* (wireless) multimedia as a potential huge success (killer application cocktail). In the somewhat related issue, one should be also careful on estimating the real short-to-mid-term value and potential for new revenues from QoS – many of the current networking and wireless problems or business models issues are not really QoS centric. The present argumentation for that seems to be relatively weak, apart of some clear niche applications. As a small note the relative success of gaming and adult content over mobile should not have become as surprise, and it would be interesting to see some reliable and recent global statistics on the mobile services distribution.

Any future network architecture, whether wireless or fixed, should take into account the tussle space. More specifically we should recognize the different requirements, and leave out enough space for dynamic solutions and tussles. Moreover, the economic aspects must be taken in account - naturally not by trying to do a vertical integration, but nevertheless economical incentives must be there. Otherwise any large-scale proposal would be failing in the real world deployment phase. The requirement to rethink architecture and protocols has risen both in wireless and Internet domain, but from the different perspectives. In the Internet it is driven by the fact that it has become multi-player commercial network. In the case of wireless, the driver has been the success of wireless access and its penetration from voice-only domain to data communications and pervasive applications.

Odlyzko[12] has also introduced some highly interesting thoughts on price discrimination in the Internet context. Following that road we are presenting some questions for likely trends also in the wireless. Our thoughts are in part same as presented also by Wolisz[13] and Liver & Dermler[14]. The crucial question we would like to ask is why do we not have more *"service provider pays for communications"* (or 1-800) type of service models. Instead of the current state of the art that seems to aim at to increase the use of user and IP-address based direct billing and QoS.

One could argue that in some cases more simplified model with possibility for higher price discrimination could be introduced through "reverse billing logic". For example think the situation, where end-user is connecting to some shop (service) to order something (video clip, book etc.). Instead of user negotiating service level agreement (which is any case a very abstract concept for him or her) and paying for connectivity (often blindly), an alternative solution would be to provide "free" and even anonymous flat initial rate connectivity to end-user. The user would buy some service from a provider (application, video clip, eCommerce, etc.). The provider sets price for wholesome service, with possibility for high price discrimination using any possible knowledge on the customer (and superior knowledge on the requirements of the service). The key point is that, if the service provider is paying (negotiating) for connectivity price and SLA (if some specific QoS is required) it has more bargaining power due to sheer volume of traffic and superior knowledge on the application and users. This volume and bargaining power could even encourage upgrades, investments and innovations in the network. Obviously this model is not universal, but the mix of the present day wireless billing and alternatives could become quite interesting. In the any case, expect to see that even more innovative billing mechanisms and more in depth studies on the economical-technological implications will become trend-in-making.

2.4 Sensor Sea

The advent of sensor networking will bring forth the need for more fundamental architecture work. When we will be surrounded by a "sea of wireless sensors", new methods of abstracting and localizing the communication complexity have to be developed. The limited resources available at the sensor nodes make it very unlikely that they will be connected to the Internet directly. Instead, some form of gateway-based architecture will most likely be used. Our suggestion on how the relationships of the future Internet and the sensor sea might look like can be summarized by the following figure (adapted from Mähönen et al.[1]): In figure 1, **I** denotes basic Internet domain, and GW means gateway functionality. The hierarchy from general interconnecting Internet goes down to different subdomains (wireless and fixed). The lowest level is **S**, which is a pervasive sea of sensors. Most sensors do not have any direct connectivity to internet and they have to rely to support from other systems.

Naturally this kind of hierarchical architecture will create difficulties when trying to understand the relation of the different levels in the hierarchy depicted in figure 1, and the different planes and layers of the communication protocol stacks. Much of the functionality of future networks will be distributed amongst the sensors and gateway nodes, and many of the more complex functions, such as context sensitive inference will most likely be handled in devices "above" the sea of sensors, only relying on the small devices for supplying the necessary information without heavy preprocessing.

The above is a definite trend-in-making which has gained already a large momentum (after years of preparation and visioning), i.e. sensor networks and network capability in multitude of embedded devices (embedded networking). This trend will give also a rise to some very though privacy and community policy issues (code of conduct, ethics on using information) – these issues are already encountered, e.g. with camera phones and wireless surveillance cameras, but in the smaller scale.

The high differentiation and hierarchy might move us back towards *"regional architectures"*. In fact, we already have some regionalism (even in IP delivery) since wireless systems and core Internet have different solutions and they are in part connected through gateways ("isolators"). Although, the current trends have been towards greater homogeneity and all IP, it is not a trivial conclusion on the long term basis. One could definitely entertain the possibility to design architecture that would use regionally and context based architectures for good effect, e.g. sensors are forcing us to accept it in any case. In the long term we should look not only extended IP networks, but also possibilities for new architectures including non IP possibilities. This leads us to move to our final topic on fully cognitive entities instead of "playing" up with context sensitivity.

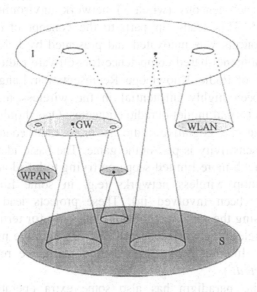

Figure 1 Hierarchy between different domains in the future networking (adapted from Mähönen *et al.*).

3. COGNITIVE ENTITIES

3.1 Adaptivity and Cognitive Radios

Adaptivity has been an integral part of wireless design for a very long period of time. This is, of course, a necessity due to inherent unreliability and variability of wireless channels. Over a decade the adaptivity boundary has been pushed from the physical layer toward the whole systems, and one of the key aims of this (re)configurability approach has been software radio. For a very large part the "classical" software radio and adaptivity in general are confined to algorithmic approach. In some areas the algorithmic approach is the only possible way to proceed at least at the lowest abstraction level, due to e.g. real-time and design constraints. However, as the complexity increases and we enter to tussle space it becomes difficult, and perhaps even downright impossible, to build algorithmic adaptivity.

Especially the struggle between providing virtual zeroconfiguration and ease-of-use for ordinary users, and making more adaptive and powerful systems in the heterogeneous (wireless) network environment is a very difficult situation. This leads, in part, to the concept of *cognitive radio paradigm*. This concept was innovated and presented by J. Mitola[6] in 2000. Mitola presented a model-based competence for software radios, and defined a early prototype of Radio Knowledge Representation Language (RKRL). This work has been highly influential in the wireless research area by introducing also a formal model of radio etiquettes. One should be careful on understanding that the cognitive radio is more than context sensitivity, although context sensitivity is part of the game. The basic idea of RKRL has been used with much more limited scope on trying to build adaptive control planes for multihop wireless networks (e.g. in some European Union projects we have been involved in). These projects lead our group to thinking about using the same methodology not only for terminals, but also as an overall (wireless) network concept (for full cognitive networking idea directly from the Internet network research domain, we refer to seminal paper by Clarke *et al.*[5]).

Cognitive radio paradigm has also some extra operational benefits. Although, if taken to its extreme it is more futuristic than software radio, it can also be implemented in partial form without building the ultimate software radio hardware. There are still formidable hardware and algorithm development problems (such as AD/DA-converters see e.g. Jamin *et al.*[10] for a simple discussion) before full (ideal) all-in-one software radio can be built. So although, software radio and cognitive radio are interlinked and are "close relatives", they have also distinctive technology roadmaps and characteristics. In short the basic paradigm in the cognitive radios is to provide technologies, which enable radio to reason about its resources, constraints, and be aware of users/operators' requirements and context. There cognitive radio research definitely is closely related to ubiquitous and pervasive computing R&D. Due to all this there is already some interesting movement in the cognitive radio domain and this field definitely merits the characterization as trend in making.

One of the emerging discussion topics seems to be using cognitive radio paradigm for dynamic spectrum allocation and use. This is in our opinion a domain, where there might be highly different expectations and goals. It is relatively easy to see how, e.g. tactical military or emergency scenarios would be very tempting cases for dynamic and cognitive spectrum handling. In those cases, finding the consensus and negotiating tussles for dynamic frequency management could be possible. For civilian (cellular and wireless) networking in short-to-mid term one tends to be more cautious (even pessimistic) on predicting possibilities. The established spectrum use paradigm is deeply rooted – also in to business and government tradition. The frequency domain is still scarce for wireless communications even though some observers are predicting the imminent dawn of "infinite bandwidth".

3.2 Cognitive Networks and Wireless

Cognitive Networks have been proposed as a possible new architectural principle for networks beyond the present day Internet[5]. They can also be seen as a logical generalization of cognitive radios (or vice-versa, cognitive radios are logical subset of idea space rising from the cognitive networks paradigm). At this time it is perhaps good time to recall the Oxford English Dictionary[16] definition for *"cognition:* knowing, perceiving, or conceiving as an act or faculty distinct from emotion and abolition; a result of this; a perception, sensation, notion or intuition." More operationally cognitive radios and networks are able to (in limited sense) reason based on appropriately represented knowledge. Apparently this should also mean that those cognitive systems have ability to learn and explain themselves (see also Brachman[11]).

The idea of the *knowledge plane* (KP) that would be separate from data and control plane is a very strong paradigm[5]. The existence of knowledge plane would enable cognitive networks and radios so that the systems would be able to answer questions like WHY(this network does not work) and ultimately do some self-configuration, provide suggestions on how to fix problems and in the more limited area self-heal and fix user problems. One of my personal favorites is also a scenario, where cognitive network can infer automatically that massive spamming operation is starting. It responds by stopping connection, and alarming both ISP and prosecutor on the possible violation. In our opinion KP design and whole cognitive network should take into account wireless networks and telephony from the beginning. There are some specific problems emerging from the wireless domain, and also wireless communications could gain tremendously from the cognitive networking, thus providing one of the economical stimuli to deploy cognitive technologies in the first place.

There are many challenges that should be taken in to account. Like in the case of fixed networks, there is a large difference between users and their technical capacities are highly different. Even the simple questions WHY (my_phone; does not work with this $game and

$network_setup) can be expressed with very different contexts. Moreover providing answers (or suggestions) to questions is even more difficult, since an answer that is perfectly acceptable for one user might be complete jargon for another. This leads to necessity to provide context and user sensitive semantics for different functionalities. The small wireless devices (especially emerging home applications and sensors) are very limited with their computational capabilities and power consumption is always an issue. How to provide some limited functionality for these devices to perform even without network connection is also a problem to be solved. In the case of smallest sensors, we might decide not to even try.

The future wireless systems would benefit a lot from the automatic (re)configurability and context information that can be provided in part by KP. One of the interesting possibilities would be to integrate circuit switched voice-telephony systems and packet switched data networks more closely, and through this to provide cross-platform co-operation, optimization and better interoperability. For complex inference we are better off by using computational capability of infrastructure and this is area, where grid and opportunistic computing might become interesting possibilities.

Even "just" developing tools toward cognitive networks would be useful. Current modeling and simulation does not enable any efficient way on expressing the high level goal(s) of the operators. This is a serious modeling problem, since the current approach that simulates only technical issues within simplified operator domains is inherently limited. The capability to express efficiently (e.g. through Network Knowledge Representation Language - NKRL) high level goals and policies, could potentially open possibility to make composable networks. Knowledge plane could thus provide better interoperability for different operator domains, and could be tremendous value for efficient spectrum use.

However, the wireless networks and terminals also provide a number of challenges. The inherently unreliable and rapidly changing channels, mobility, and potentially frequently changing context are among the many challenges. Telecommunication world is also very traditional, and opening information towards open knowledge plane requires good justifications and security arguments. Any of these problems do not need to be fundamental roadblocks; on the contrary they are good challenges to make grand challenge of cognitive radios and networks even more interesting. The key issue is not to confuse cognitive networks and radios (CNR) with context sensitivity or pervasive computing. The applications, service discovery, knowledge presentation etc. are attracting a lot of research on these areas, and will provide good stepping-stones, but CNR domain is something larger. Obviously issues on cognitive framework and architecture are important, especially providing *machine learning* and *AI* support to networks, also defining suitable Knowledge Representation Languages, Protocols and Infrastructures are important. Our group's own early approach has been to consider the use of (Kohonen's) Self-Organizing Maps[18] as self-learning clustering and automated data exploring tools for CNRs (work in progress).

Although the issue of providing KP and composable networks is alone a formidable task, there is also a number of subtasks to be solved more urgently. Good example is how to provide more direct information about link layer(s), link conditions, network capabilities etc. as transparently as possible to control planes and KP. Although some of this can be seen as traffic over data plane and some sort of signalling, there is most probable necessity to define new (distributed) APIs and socket interfaces in order to enable to access to information. Especially some applications, system programmers and operating systems would benefit on this.

Naturally there is a long research road ahead before even basic context sensitivity can be integrated into the networks. Problems in different inference techniques have been proven surprisingly difficult to solve, as has been shown by the critique aimed at various "helper functions" that have been introduced into many commonly used (desktop) computer programs. In the networking use the even larger heterogeneity of the environment and that of the users will make these problems even more acute. Note also how difficult it seems to be to do even basic context sensitive applications[17].

3.3 Is it alive and can I play with it?

As a final remark, it is interesting to note that game theory has been started to apply for communications problems with increasing frequency – with very good results. Roughly speaking the game theoretical approach has become stronger from two different perspectives. First, just like in the economics, wherein game theory is an established tool, many problems in communications require understanding of 'economically' based behavior. Second, the co-operation is required at different level between players, and thus game theory is quite natural tool. Especially in the case of large networks and *adhoc* networks there is necessity to model behavior and incentives for co-operation. I would argue that these methods are becoming more and more mainstream in coming years.

An intriguing idea is that as the networks become more complex and larger with some components of adhoc networking, they start to resemble not only physical objects, but also behavior that are similar to biological and large economical systems. This idea is not a new, but requires probably some rethinking. As we are now strongly involved in to help computational biology, perhaps in the future we should consider also the use of their modeling methods towards networking problems. In the case of co-operative and adhoc networks it seems that extending the game theory to include evolutionary aspects and high dynamic behavior become a very strong tool for many (wireless) networking problems (see some introduction to game approach by Axelrod[21] and Hofbauer[22]). Urpi *et al.*[23] have proposed a model of cooperation based on Bayesian games, and Félegyházi *et al.* [24] have studied packet forwarding strategies just to mention few recent works.

More work towards understanding the underlying mechanism of dynamic behavior systems of the systems is required. The current trend is focused very much to adhoc networks, but it is to be expected that same tools will be highly useful also for other type of large networks, and one can speculate that especially research on understanding trust and behavior in Internet could gain from using evolutionary biology, immunology, and game theory tools. Overall, it is tempting to see different cooperative network scenarios as ecosystems, which can be studied and modeled using game theory and computational biology. More boldly if the future networks have scale-free, small world connectivity and total number of nodes are on the order of 10^{12} – 10^{15}, should we ask; "is it alive?" (see also the discussion by Steele[2]).

4.   CONCLUSIONS AND OTHER TRENDS

The above discussion argues that different machine learning and intelligent methods are most probably becoming more ubiquitous and pervasive up to the point that we might want to change *both* our core-network and wireless architectures to support such concepts.

Tied in to development of wireless and fixed communications are also other interesting possible trends. One possible trend is the emergence of *"GreenChips"* approach. With GreenChips we mean that overall environmental trend might catch us. If so in the future we would try to minimize the number of large antennae (esthetics), lower power consumption (good for many reasons, but also for energy saving), and overall build GreenChips that would be environmentally friendly. Although, the debate of the total power consumption of Internet and Computer might be an interesting issue[19], we do not mean that. The power consumption issue is, obviously, important for wireless systems for operational reasons. And GreenChip approach might become even extremely necessary, if we want to introduce a *throwaway* concept for sensors. The new materials developed in laboratories could make GreenChips even compostable; the biologically friendly solutions could be thus very valuable for both sensor and medical applications. As a thought from Science Fiction, but not without previous serious thinking, cf. Maguire[20] and McGee & Steele[2], is to speculate that the current basic research on brain signal scanning and control will make a leap towards actual commercial technology and social acceptance for systems increases – a truly cognitive network and radio would be interacting with our brain. By that though we have moved from the future research visioning to more science fiction, but at least this is a long-term vision for telecommunication that has been sometimes asked.

In the any case, we can say safely that we are still living *"interesting times"* whether we are working in the fixed or wireless networking. The emergence of ever complex system and sea of sensors is leading to networked systems that have a biological complexity and behave probably in part like statistical (physical) systems. Raymond Steele[2] was asking, if we know what we are doing and maybe we are building a global brain. It is probably safe to say that we cannot predict the future and do not completely know what we are building out -- but outcome is going to be interesting.

ACKNOWLEDGEMENTS

Author thanks several collaborators and friends who have contributed to though process and corrected numerous early misconceptions. Especially I am acknowledging help from and discussions with J. Pereira, M. Petrova, J. Riihijärvi,C. G. Maguire, A. Wolisz and K. Wrona. Moreover Prof. Wolisz made my thinking on pricing issues much more clear, and also pointed out the references [13] and [14]. The work is part of our internal "tiny and cognitive devices" program. The work has been supported in part by grants from Academy of Finland, European Union, DFG and Ericsson Research.

REFERENCES

1. P. Mähönen, J. Riihijärvi, M. Petrova, Z. Shelby, "Hop-by-Hop Toward Future Mobile Broadband IP", *IEEE Comm. Magazine*, vol. 42, No.3., 138-146 (2004).
2. R. Steele, "Full-Ahead to Where", *ECWT 2000*, Paris, France 2000. See also R. Steele, "Communications++: Do We Know What We Are Creating", *Proc. EPMCC 1997*, Berlin, Germany, 19-23 (1997).

3. D.-J. Farber, "Predicting the Unpredictable: Future Directions in Internetworking", *IEEE Comm. Magazine,* vol. 40, no. 7, 67-71 (July 2002).
4. NSF, "Report of the National Science Foundation Workshop on Fundamental Research in Networking", NSF, CISE, (April 2003).
5. D. D. Clark, C. Partridge, J.C. Ramming, J. T. Wroclawski, "A Knowledge Plane for the Internet", *ACM SIGCOMM 2003,* Karlsruhe, Germany, (2003).
6. J. Mitola, Cognitive Radio, Ph.D. thesis, KTH, Stockholm, (2000).
7. D. Clark, J. Wroclawski, K.R. Sollins, R. Braden, "Tussle in Cyberspace: Defining Tomorrow's Internet", *ACM SIGCOMM* 2002, Pittsburgh, PA, USA, (August 2002).
8. A. M. Odlyzko, "Content is not king", *First Monday,* 6(2) (February 2001), http://firstmonday.org/.
9. A. M. Odlyzko, "Internet traffic growth: Sources and implications", *Proc. Optical Transmission Systems and Equipment for WDM Networking II,* B. B. Dingel, W. Weiershausen, A. K. Dutta, and K.-I. Sato, eds., SPIE, vol. 5247, 1-15 (2003).
10. A. Jamin, P. Mahonen and Z. Shelby, "Software Radio Implementability of WLANs", in *Proc. 12th Thyrrenian International Workshop On Digital Communications* - Software Radio Technologies and Services, Porto Ferraio - Island of Elba, Italy, 13-16 (2000).
11. R. J. Brachman, "Systems That Know What They're Doing", *IEEE Intelligent Systems,* pp. 67-71 (2001).
12. A. M. Odlyzko, "Privacy, economics, and price discrimination on the Internet", *Proc. of ICEC2003; Fifth International Conference on Electronic Commerce,* N. Sadeh, ed., ACM, 355-366 (2003)
13. A. Wolisz, "Information access is fine, but who is going to pay?" in *New Developments in Distributed Applications and Interoperable Systems,* (K. Zielinski, K. Geihs, A. Laurentowski (eds)), Kluwer, 149-160 (2001).
14. B. Liver, G. Dermler, "The E-Business of Content Delivery", Workshop on Internet Service Quality Economics, MIT, Dec. 2-3 (1999).
15. J.M. Kahn, R.H. Katz, K.S.J. Pister, "Next Century of Challenges: Mobile Networking for Smart Dust", *Proc. ACM Mobicom 1999,* 271-78 (1999).
16. The Concise Oxford English Dictionary, Oxford, Oxford University Press (2003).
17. T. Erickson, "Some Problems with the Notion of Contex-Aware Computing", *ACM Commun.,* vol. 45, no. 2., 102-104 (2002).
18. T. Kohonen, Self-organizing maps, Springer-Verlag, New York, (1997).
19. T. S. Perry, Fueling the Internet, *IEEE Spectrum,* January 2001, pp. 80-81 (2001).
20. E. McGee and G. Q. Maguire Jr., "Ethical Assessment of Implantable Brain Chips", *Proceedings of the Twentieth World Congress of Philosophy,* August, (1998).
21. R. Axelrod, The Evolution of Cooperation, Basic Books, New York, (1984).
22. J. Hofbauer, and K. Sigmund, Evolutionary Games and Population Dynamics, Cambridge University Press (1998).
23. A.Urpi, M. Bonuccelli, and S. Giordano, Modelling Cooperation in Mobile Adhoc Networks: A Formal Description of Selfishness', *Proc of the Workshop on Modeling and Optimization in Mobile, Adhoc and Wireless Networks.* INRIA, Sophia-Antipolis, France, (2003).
24. M. Félegyházi, L. Buttyáan, J.-P. Hubaux, Equilibrium Analysis of Packet Forwarding Strategies in Wireless Adhoc Networks – the Static Case?, *Proc. of Personal Wireless Communications (PWC 2003),* Venice, Italy, Sep 23-25, 2003, Lecture Notes of Computer Science, Volume 2775, Springer-Verlag (2003).

Chapter 4

MULTIPLE ANTENNA SYSTEMS: FRONTIER OF WIRELESS ACCESS

E. DEL RE, L. PIERUCCI

University of Florence, Department of Electronics and Telecommunications, S.Marta 3, 50139 Firenze, Italy

Abstract: Multiple antenna systems are the new frontier for wireless communications including the actually third generation mobile communication systems, called Universal Mobile Telecommunication System (UMTS), the wireless LAN and the wireless PAN up to the future 4G mobile system focused on seamlessly integration of the existing wireless technologies. The use of multiple antenna systems improves the overall system performance in term of capacity and spectrum efficiency achieving high data rate wireless services. The main multiple antenna processing techniques are highlighted in terms of performance /complexity tradeoffs: smart antennas with adaptive beamforming to cancel the interference signals (from other users or multipath) and MIMO systems to exploit the space-time properties of wireless channels.

1. INTRODUCTION

Recently, in wireless communication systems great interest has been shown for the use of multiple antennas to increase coverage and spectral efficiency.

The popularity of GSM led to a typical site with 120° sectorized hardware beamforming antenna array increasing the spectrum sharing. The direction and gain of sectorized antennas are fixed and to each beam a distinct channel is assigned.

When the requirements of capacity are exceeded, smart antennas with adaptive digital signal processing are necessary. The smart antenna's beams are not fixed but are dynamic for user and will place nulls in the radiation pattern that cancel interferences. The application of adaptive antenna techniques to fixed-architecture base stations has been shown to offer wide-ranging benefits including interference rejection capabilities. In particular, the antenna array should allow the identification of the mobile desired user by using proper DOA (Direction of Arrival) estimation algorithms and improve the signal-to-interference-plus- noise ratio steering the main beam in downlink according the above direction estimate.

In the third generation mobile systems, called Universal Mobile Telecommunication System-UMTS, based on CDMA access, the use of multiple antennas allows the spatial separation of user signals, decreasing the multiple access interference (MAI) effect. Therefore, space-time processing can significantly increase the capacity of CDMA systems. Diversity combining schemes are proposed such as the Matched Desired Impulse Response (MDIR) approach which performs a spatial cancellation following by a Rake receiver. Joint detection of all bits from all users can overcome the limitations of MAI. In advanced multiuser beamforming approach, smart antennas are combined with interference-canceling multiuser receivers showing a greater performance for the uplink CDMA systems [1].

In UMTS system the use of quasi-orthogonal codes and the multipath of the wireless channel result in a critical MAI [2].

In rich scattering environments the system capacity can be increased by exploiting decorrelation of transmitted signals [1]. In theory, the capacity increases proportionally with the minimum number of antennas at the transmitter or at the receiver. Space-time coding and modulation, and spatial multiplexing MIMO systems exploit different independent parallel transmission channels to provide high data rates at no cost in bandwidth or power.

Each multipath reflector is not seen as a drawback to be equalized out, but can be seen as an independent path for independent bit streams.

Lucent decides to beamform into multipath and their Bell Laboratories Layered Space-Time (BLAST) approach [3] [4] shows multipath channel capacity greater than that of a merely noisy channel where, according the Shannon law, the capacity is a function of bandwidth and signal-to-noise ratio.

However, the requirement of rich channel scattering is critical and in realistic environment channel coding is needed to approach the MIMO system capacity.

The major source of channel impairment in a spatial multiplexing scheme is co-antenna interference (CAI). To mitigate the degrading effects of CAI, a robust multi-transmit multi-receive system using the combination of Turbo decoding principles and V-BLAST, was analyzed in literature [5], [6], called T-BLAST.

The paper first describes smart antenna systems reviewing the concept of adaptive bemforming and DOA estimation to improve the signal-to-interference ratio, highlighting the advantages when they are used in CDMA systems. In the third section an example of DOA estimation particularly optimized for the UMTS environment is considered [7].

Section IV reviews the concepts of MIMO systems: space-time coded modulation and the iterative space-time layered multiple antenna receiver of the Bell Laboratories. The last section shows an example of MIMO system, the combination of the BLAST architecture and the iterative demapping and decoding of the Turbo decoding principle. The performance of this T-BLAST structure was evaluated according to the IEEE 802.11b standard requirements for Wireless LAN. Finally, future challenges and issues of MIMO systems are presented.

2. SMART ANTENNAS

In the SDMA (Space Division Multiple Access) the capacity of the mobile cellular system increases exploiting the 'spatial diversity' of the users. In particular the radiation pattern of the antenna at the base station should be dynamically modified steering the main beam toward the desired user and by placing the nulls in the pattern corresponding to the interference directions (null steering method). In a personal communication system, this can be achieved using antenna array and an adaptive system is required in order to shape the radiation pattern of the array. The antenna array provides a greater receiving gain than an omnidirectional antenna allowing to transmit with a lower power level and to reduce electromagnetic pollution levels.

The SDMA techniques increase the overall performance of the radio mobile system in terms of number of users simultaneously accepted in the same channel. This spatial orthogonality allows to separate the signals from different terminal users even if they are overlapped in the frequency/time domain (FDMA/TDMA techniques) increasing the spectral efficiency. In the case of CDMA access where all the users share the same channel and each user has a different pseudonoise code, the use of smart antennas improves the signal-to-interference ratios allowing reducing the MAI.

2.1 Adaptive Beamforming

In a space diversity system the weighted sum of the received signals is combined at the output of beamformer to steer a beam toward the user signal and adjust the nulls to reject interferences. Different choices of algorithms are possible with a different degree of complexity:

- Based on DOA estimation: if the direction of the desired user is known, the beamformer tries to minimize the output power constrained to maintain a distortionless response in the direction of interest (Minimum Variance Distortionless Response-MVDR). This method depends strongly on spatial information knowledge (e.g. MUSIC,ESPRIT), is computationally expensive, and the number of users is limited by the number of antenna elements
- Based on training sequence: the beamformer employs a training sequence to minimize the difference between the training sequence and its output in the mean-square sense (MMSE). The use of a training sequence reduces the capacity of the system.
- Blind methods (Constant Modulus-CMA): It works on the premise that the presence of interferences and multipath causes distortion on the amplitude of array output which has a constant modulus. It is only effective for constant modulated envelope signals such as GMSK and QPSK.

2.2 Space-time processing in CDMA

In the uplink of a wideband direct sequence code-division multiple access (W-CDMA) system the orthogonality of the variable spreading factor (OVSF) channelization codes and scrambling codes (Gold sequence) for each user is affected by severe multipath interferences especially for high-rate users. Therefore, by employing the adaptive antenna array transmitter diversity we can increase the forward link capacity suppressing the MAI. The transmitter antenna weights are modified according the antenna weights generated in the coherent adaptive antenna array diversity receiver.

The downlink weight vector was calculated using uplink measurements, assuming that uplink and downlink frequencies are sufficiently closed to disregard the discrepancies between responses.

A single-user space- time RAKE receiver consists in digital matched filters, a digital beamformer (MMSE type) matched to each path followed by a standard Rake used to phase recombine the delayed signal components. Taking into account the fact that different paths come from different directions it is possible to resolve even path which are irresolvable in the time, due to the low delay spread. The receiver antenna weights are adaptively controlled so that the mean square error between the Rake output signal and the reference signal is minimized using an adaptive algorithm (e.g. LMS,RLS) as shown in figure 1 [8].

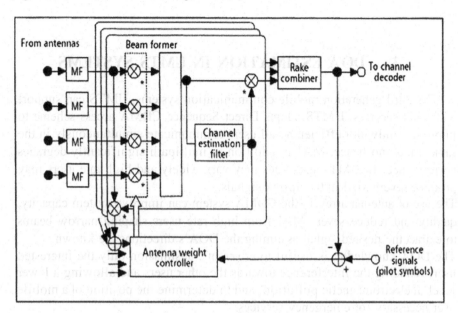

Figure 1 Coherent adaptive antenna array MMSE receiver

In a WCDMA system the conventional matched filter (MF) receiver detects the user's data due to the known spreading code considering the other users signals as noise.

Jointly detection of all data from all users exploits the structure of MAI (weaker and strong users). Among the many multiuser detectors, interference cancellation techniques are relatively computationally efficient.

The Successive Interference Cancellation (SIC) sequentially removes the MAI due to the stronger users before detecting the weaker ones. Parallel Interference Cancellation (PIC) receivers simultaneously cancel all interfering users from the user of interest, before taking the final symbol decision.

In Selective-Parallel Interference Cancellation the output of a bank of matched filters (one for each signal replica) and Rake is compared with a suitable threshold and assumed reliable or not, depending on its value.

The reliable users are directly detected and cancelled from the whole received signal before making the decision on unreliable signals or replicas. Therefore, the MAI effects of reliable signals (the strongest) on the other signals are cancelled at this second step [9].

In particular multiuser beamforming can decrease the MAI effects combining adaptive antenna array with the interference–cancelling multiuser receiver for the WCDMA uplink [9]. However hardware complexity increases according to the number of accessing users.

3. DOA ESTIMATION IN UMTS SYSTEMS

The third generation mobile communication systems, UMTS can support wideband services. UMTS adopts Direct Sequence CDMA access scheme to improve bandwidth efficiency, all users communicate simultaneously in the same band and hence, MAI in addition to multipath significantly degrades performance. In UMTS data rates may vary widely and high rate users may produce severe MAI in the uplink signals.

The use of antenna array in the CDMA system can improve system capacity, quality and reduce severe MAI from high rate users shaping narrow beams to extract the desired signal assuming the DOA's directions are known.

The DOA knowledge is useful to cover in transmission only the interested user, reducing the interference towards the other users and allowing a lower level of electromagnetic pollution, and to determine the position of a mobile user necessary for emergency services.

Conventional DOA methods using the eigenvectors decomposition are based on the assumption that the number of antenna elements is greater than the number of impinging signals (users and replicas). Therefore in multiuser environment with multipath channel these algorithms are not suitable. In CDMA system each user is identified by a unique code orthogonal to the other user codes. Therefore in an optimized approach for UMTS system a DOA estimator (MUSIC or Iterative-MUSIC) is implemented for each user for each replica [7]. The estimators work at the output of matched filters, one for each path. In this case only one peak of MUSIC spectrum is estimated at a time, providing a high reliability in DOA estimate values and it is more independent from the effect of multiple paths even if the complexity increases.

Unfortunately, the scrambling codes proposed in the 3rd Generation Partnership Project (3GPP) standards have not orthogonality properties to guarantee adequate user identification and the multipath channel contributes to degrade the codes properties. Therefore, the uncancelled MAI decreases the performance of DOA algorithm. However, the simulation results (in Figure 2) show a remarkable behaviour in DOA estimation also in the case of 64 users configuration for a global rate of 3.78Mb/s (full chip rate is 3.84 Mchips/s, according the 3GPP standard).

4. MIMO SYSTEMS

Multiple antennas at the transmitter and receiver provide diversity in a fading environment. By employing multiple antennas, multiple spatial channels are created, and it is unlikely the entire channel will fade simultaneously. If statistical decorrelation among antenna elements is provided, this condition can be satisfied by using antennas well separated (by more than $\lambda/2$) or with different polarization, multiple transmit and receive antennas can create independent parallel channels. In rich scattering environments the system capacity can be increased exploiting decorrelation of transmitted signals by using the space-time coded modulation (STCM) and processing of MIMO systems.

Especially in 3GPP downlink the use of multiple transmit antennas using suitable space-time code modulation can achieve the required high data rates.

Various transmit diversity techniques are proposed in literature: the delay transmit scheme by Seshadri and Winters [12] where the same signal is transmitted by multiple antennas using different delays; the space-time block coding (STBC) by Alamouti [15] where a pair of symbols and their transformed version are transmitted by two antennas allowing the use of ML detection with linear effort and finally, the space-time coded modulation (STCM).

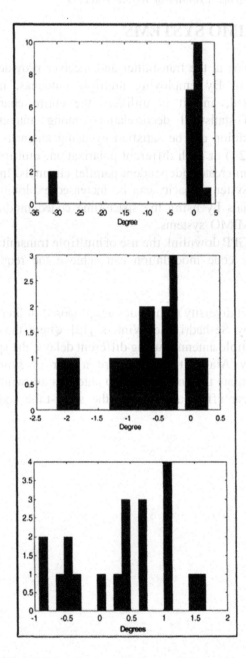

Figure 2 Iterative-Music DOA estimation - Histogram of error estimation around the DOA true value of desired user and Mean square error in a 64 users configuration for three main replicas of a multipath channel with 6 paths – Path 1(DOA=35.9° MSE=100.93) - Path 2 (DOA=-30.6°MSE=0.98) - Path 3 (DOA=2.4°MSE=0.68)

Generally, in the STCM approach data is encoded by a space-time (ST) channel encoder and the output is split into N streams to be simultaneously transmitted using N transmit antennas in orthogonal manner. The received signal at each antenna is a linear superposition of the transmitted signals perturbed by flat fading channel and noise. This coding across spatial channels uses ideas similar to Trellis Coded Modulation (TCM) to define mappings of symbols to antennas. The space-time decoding process at the receiver requires trellis decoding and may have high complexity.

Designed for simultaneous diversity and coding gains, different proposals for STCMs in the literature are mainly distinguished based on single antenna coded modulations, i.e., TCM, multiple TCM, multilevel coded modulation and BICM that they were extended from. Owing to its integrated trellis coding and higher order modulation design with the redundancy expansion in both signal and antenna space, space-time trellis codes (STTCs) [10], also referred to as ST TCM, can be seen as a generalization of the TCM to multi-antenna systems. STTCs evolved from the combined transmit delay diversity (DD) and receive MLD, originally introduced for two transmit antennas in [11,12] and further generalized to an arbitrary number of transmit antennas in [13, 14].

The Bell Laboratories Layered Space-Time (BLAST) approach achieves spectral efficiencies as high as 40 bits/sec/Hz in short-range, rich scattering wireless environment for a fixed total transmit power.

Vertical–BLAST (VBLAST) relies on spatial multiplexing at the transmit multiple antennas and spatial filtering at the receive multiple antennas.

The basic idea is to transmit different signals simultaneously on different antennas and this spatial diversity relies on a rich scattering matrix channel.

V-Blast rather than jointly decoding the signals from all the transmit antenna realizes a beamforming in multipath and first decodes the strongest signal, than cancels the effect of this transmit signal from each of the received signals and then proceeds to decode the strongest second signal of the remaining transmit signals. It is evident the analogy with the multiuser sequentially interference cancellation. The drawback of this method can be envisaged in the order for the sequential estimation and cancellation, from the strongest signal to the weakest one, and in the propagation of errors from one step of detection to the next [4].

5. T-BLAST ARCHITECTURE

To mitigate the degrading effects of CAI, a robust multi-transmit multi-receive system using the combination of Turbo decoding principles and V-BLAST was analyzed in literature [5], [6], called TBLAST.

In this section, the use of the simplified iterative interference receiver as in [5], and its performance in the specific context of the IEEE 802.11b standard is underlined.

The IEEE 802.11b standard adopts high data rate with bit rates up to 11 Mbps. For achieving data rate greater than 2 Mbps, the IEEE 802.11b standard specifies the Complementary Code Keying (CCK) modulation scheme.

The IEEE 802.11b complementary spreading codes have code length 8 and a chip rate of 11 Mchip/s. The 8 complex chips comprise a single symbol. By making the symbol rate 1.375 Msps, the 11 Mbps waveform ends up occupying the same approximate bandwidth as that for the 2 Mbps 802.11b QPSK waveform. Other IEEE 802.11b requirements are:

- It operates in 2.4GHz–2.4835GHz frequency band
- Direct Sequence Spread Spectrum (DSSS)
- Three non–overlapping 22 MHz channels

In the T-BLAST architecture a single data stream is demultiplexed into M substreams, where M is the number of transmitting antennas, and each substream is then encoded into symbols and fed to its respective transmitter. Transmitters 1-M operate in co-channel way at symbol rate with synchronized symbol timing.

It is assumed that the same constellation is used for each substream and that transmissions are organized into bursts of L symbols. The power radiated by each transmitter is proportional to 1/M, so that the total radiated power is constant and independent of M. A straight forward way to implement coding for V-BLAST is to use the *Horizontal Coding* architecture shown in Figure 3.

Each layer is encoded separately. In the horizontal encoder operation each layer uses a Parallel Concatenated Convolutionally Coded (PCCC) Turbo code, but other error correcting codes can also be adopted.

The optimal receiver processing for a coded BLAST system requires a global ML (Maximum Likelihood) solution, which jointly considers the detection (demodulation), deinterleaving and decoding of error correcting codes. However, due to the high complexity of such an approach many suboptimal techniques have been considered, including linear processing techniques such as *Zero–Forcing (ZF)* or *Minimum Mean Square Error*

(MMSE) method, and non–linear methods such as *Ordered Successive Interference Cancellation (OSIC)* [6].

The performance of these suboptimal detection methods may be improved through iterative detection and decoding, which utilize the decoding results for a second step of processing.

To extract the desired signal, Zero Forcing (or MMSE)–OSIC method at the first step and interference cancellation and Maximal Ratio Combining (MRC) in a layer-by-layer fashion at the subsequent steps are implemented.

After the first iteration, where the receiver can start V-BLAST detection and decoding from any layer, depending on the observed channel matrix, interference cancellation is performed in a layered-by-layered fashion using the present soft decoding decisions for the already decoded sublayers and the previous iteration decisions for the others.

The parallel Soft-Input-Soft-Output decoders provide the a priori probabilities of the transmitted substreams by using the bit-by-bit MAP decoding method.

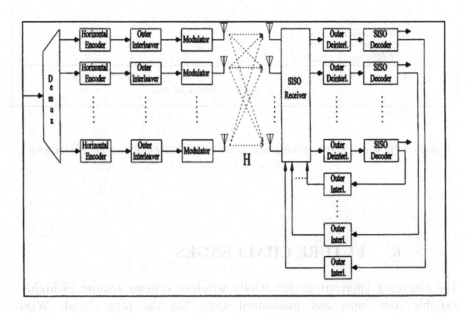

Figure 3 Horizontal Coding Turbo-BLAST Scheme.

The Figure 4 shows the performance gain of T-BLAST in term of FER versus E_b/N_0 increasing the number of receive antennas in a flat fading channel using 4 transmit antennas, DQPSK modulation and a data rate of 2 Mbps [6].

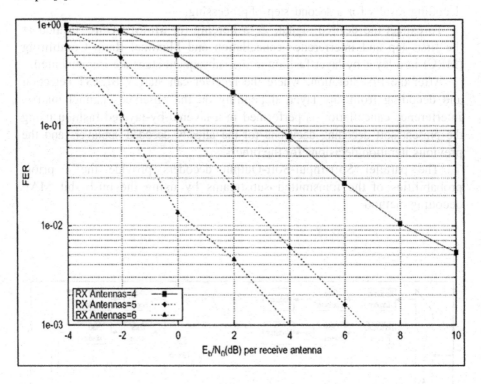

Figure 4 FER versus Eb/N0 of Horizontal Turbo-BLAST-ZF-OSIC respectively for 4,5,6 Receive Antennas

6.   FUTURE CHALLENGES

The emerging applications for 3G/4G wireless systems require of highly variable data rates and guaranteed QoS. On the other hand, WiFi technologies (and in general WLAN and WPAN services) offers higher speed in a small area. A new technology is necessary to provide a seamlessly integration of the existing wireless technologies and cellular systems. To achieve high capacity and spectrum efficiency MIMO systems will play fundamental role. Different problems have to be yet addressed:

- The layered space-time architecture requires synchronized symbol timing and the exact knowledge of the channel to achieve the promised high spectral efficiency
- Multiple antenna deployments requires multiple RF chains typically very expensive. Optimal antenna subset selection can be a promising solution.

New research should cover the analysis of the overall benefits of MIMO systems in real-world wireless scenarios (for the actual and the future services), the study of new space-time coding for multiuser environments, the analysis of new turbo codes techniques to obtain a user terminal at low cost, low power consumption supporting for multiple standards.

REFERENCES

1. S. Blostein, H. Leib "Multiple Antenna Systems: their role and impact in future wireless access" *IEEE Communications Magazine,* July 2003
2. 3GPP Standard http://www.3gpp.org release 1999.
3. G. Golden,C.J. Foschini and R.A. Valenzuela, "Detection algorithm and initial laboratory VBLAST space time communication architecture"', *Electronics Letters,* January 1999
4. B.Hassibi, "An Efficient Square-root Algorithm for BLAST", *Bell Labs* 2000
5. M. Sellathurai and S. Haykin, "'Turbo-Blast for wireless communications: theory and experiments"', *IEEE Trans. on Signal Processing,* V.50,n.10,October 2002.
6. A.Bernacchioni, E. Del Re, R. Fantacci, L. Pierucci, "T-BLAST Architecture for the IEEE 802.11b context" *GLOBECOM 2003,* San Francisco
7. E. Del Re , L. Pierucci, S. Marapodi," On the application of DOA estimation techniques to UMTS system" *IEEE 7th Int. Symp. On Spread Spectrum Tech. & Appli.,* Praha, Sept. 2-5, 2002
8. H.Taoka et al, "Adaptive antenna array transmit diversity in FDD forward link for WCDMA and broadband packet wireless access" *IEEE Communications Magazine,* April 2002
9. Del Re E., Fantacci R., Marabissi D., Morosi S., Armani C.: "Low Complexity Selective Interference Cancellator for a WCDMA Communication System with Antenna Array", *IEEE Transactions on Vehicular Technology,* Vol-52, n. 4, pp. 1162-1166, Luglio 2003
10. Tarokh V, Seshadri N and Calderbank AR (1998) Space-time codes for high data rate wireless communication: Performance criterion and code construction. *IEEE Trans Inform Theory* 44(2) (2):744-765.
11. Wittneben A (1993) A new bandwidth efficient transmit antenna modulation diversity scheme for linear digital modulation. In: *Proc. IEEE Int. Conf. Commun.,* vol3, 1630-1634.
12. Seshadri N and Winters JH (1993) Two signaling schemes for improving the error performance of frequency-division-duplex (FDD) transmission systems using transmitter antenna diversity. In: *Proc. IEEE Veh. Technol. Conf.,* Secaucus, USA, 508-511.

13. Winters JH (1998) The diversity gain of transmit diversity in wireless systems with Rayleigh fading. IEEE Trans Veh Technol 47(1):119-123.

14. Li S, Tao X, Wang W, Zhang P and Han C (2000) Generalized delay diversity code: A simple and powerful space-time coding scheme. In: Proc. Conf. ICCT, 1697-1703.

15. S.Alamouti A simple transmit diversity technique for wireless communications *IEEE jour.on selected areas in communications* vol. 16n.8 October 1998

Chapter 5

FIXED RELAYS FOR NEXT GENERATION WIRELESS SYSTEMS

Concept, Protocol, Performance and Spectral Efficiency

NORBERT ESSELING, BERNHARD H. WALKE, RALF PABST
*Chair of Communication Networks, Aachen University (RWTH), Kopernikusstrasse 16,
52074 Aachen, Germany*

Abstract: This chapter presents a concept and the related performance evaluation for a
wireless broadband system based on fixed relay stations acting as wireless
bridges. The system is equally well-suited for both dense populated areas and
wide-area coverage as an overlay to cellular radio systems. A short
introduction is given to the general topic of fixed relaying. The proposed
extension to a Medium Access Control-frame based access protocol like
IEEE802.11e, 802.15.3, 802.16a and HIPERLAN2 is outlined. A possible
deployment scenario is introduced and the simulative traffic performance in a
Manhattan-like dense urban environment and a wide-area open-space
environment is presented. It is established that the fixed relaying concept is
well suited to substantially contribute to provide high capacity cellular
broadband radio coverage in future (NG) cellular wireless broadband systems.

1.   INTRODUCTION

Future broadband radio interface technologies and the related high
multiplexing bit rates will dramatically increase the traffic capacity of a
single Access Point (AP), so that it is deemed very unlikely that this traffic
capacity will be entirely used up by the user terminals roaming in an APs
service area. This observation will be stressed by the fact that future
broadband radio interfaces will be characterised by a very limited range due
to the very high operating frequencies (> 5 GHz) expected. Furthermore,
future broadband radio systems will suffer from a high signal attenuation
due to obstacles, leading either to an excessive amount of APs or to a high
probability that substantial parts of the service area are shadowed from its

AP. By means of traffic performance evaluation, this paper establishes that a system based on fixed mounted relay stations is well suited to overcome the problems mentioned. The paper is organised as follows. The introduction explains the advantages of relaying, presents fundamentals on how the proposed relaying concepts works in general and finally explains how to "misuse" existing standards to enable relaying in the time domain for wireless broadband systems based on a periodic Medium Access Control (MAC) frame, as used in IEEE802.11e, 802.15.3, 802.16a and HIPERLAN2 (H2). The latter system is taken to exemplify a detailed solution. Section 2 answers the question under what circumstances a relay based 2-hop transmission should be preferred to a 1-hop transmission between Mobile Terminal (MT) and AP. Section 3 presents the simulation environment, important parameters and the deployment scenarios used to obtain the performance results, which are given in Sections 4 and 5. Conclusions are drawn in Section 6.

1.1   Characteristics of the Relaying Concept

The properties of our relay concept and the benefits that can be expected are as follows:

Radio Coverage can be improved in scenarios with high shadowing (e.g. bad urban or indoor scenarios). This allows to significantly increase the Quality of Service (QoS) of users in areas heavily shadowed from an AP. **The extension of the radio range** of an AP by means of Fixed Relay Stations (FRS) allows to operate much larger cells with broadband radio coverage than with a conventional one-hop system. The FRS concept provides the **possibility of installing temporary coverage** in areas where permanent coverage is not needed (e.g. construction sites, conference-/meeting-rooms) or where a **fast initial network roll-out** has to be performed. The wireless connection of the FRS to the fixed network **substantially reduces infrastructure costs,** which in most cases are the dominant part of the roll-out and operations costs. FRS only need mains supply. In cases where no mains is available, relays could rely on solar power supply. A **standard-conformant integration of the relays into any MAC frame based system** would allow for a **stepwise enhancement of the coverage region** of an already installed system. Investments in new APs can be saved and any hardware product complying with a wireless MAC frame based standard is possible to be used without modifications. The proposed relay concept can be recursively used to extend the radio coverage range of a single AP by multi-hop links.

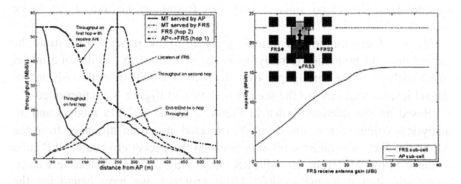

Figure 1. Left: Throughput for separate hops and end-to-end for MTs served by FRS (16dBi FRS receive antenna gain); Right: Capacity of the AP in single-hop mode and capacity of a FRS

In this case, a FRS serves another FRS according to the needs besides serving the MTs roaming in its local environment. It is worth mentioning that we focus on relaying in layer 2 by means of what is called a bridge in Local Area Networks (LANs).

1.2 Fundamentals

In relay based systems, additional radio resources are needed on the different hops on the route between AP and MT, since multiple transmissions of the data have to take place. We have studied three concepts and present here the results of the first one.

Relaying in the Time-Domain: The same frequency channel is used on both sides of the relay. A certain part of the MAC-frame capacity is dedicated to connect MT and FRS and the rest is used to connect AP and FRS via a time-multiplexing channel. One transceiver only is needed in a FRS, which results in cheap, small and energy-efficient FRSs. The physical layers of the standard air interfaces considered do not require any modifications. Instead the FRS concept is realised through the MAC protocol software only.

Relaying in the Frequency-Domain: This concept uses different carrier frequencies on links a FRS is connecting. The two hops can be operated independently of each other at the cost of increased complexity of the hardware and the frequency management.

Hybrid Time-/Frequency-Domain Relaying: In the hybrid concept[1], the FRS periodically switches between two frequencies, allowing the AP to continue using its frequency f_1 while the relay serves its terminals on frequency f_2. No additional transceiver is needed, but the hardware complexity is increased since a very fast frequency switching has to be supported.

We will focus on the time domain relaying in this paper. To illustrate the capabilities and properties of relaying in the time domain, results of a model based analysis of the throughput over distance of a MT from the AP and of the achievable capacity for the scenarios shown in Figure 8 are presented.

Based on the relation shown in Figure 5 for an 802.11a modem and an analytical calculation of the *C/ (I+N)* expected at certain distances from the AP and/or FRS, we obtain a relation between the packet error rate (PER) and distance of the MT from the AP/FRS. Assuming an ideal Selective Reject-Automatic Repeat Request (SREJ-ARQ) protocol, we have calculated the resulting relation between throughput and distance from the AP/FRS; see the solid curve in Figure 1 (left). We assume further that the FRSs have directive transmit/receive antennas to communicate with the AP and an omnidirectional antenna to communicate with MTs. Gain antennas at the FRS result in an improved throughput-distance relation between AP and FRS, as is visible from the dotted curve in Figure 1 (left). The throughput of a MT that is served by the FRS (dashed curve in Figure 1, left) in general obeys the same throughput-distance relationship that is also valid for MTs served by the AP directly. The dash-dotted curve finally denotes the maximum achievable two-hop throughput of a MT served by the FRS. It is clearly visible that a considerable extension of the radio coverage range can be achieved through the use of the relay station with a 16 dBi gain antenna assumed. Figure 1 (right) gives the capacity of the AP sub-cell (horizontal line) and compares that with the FRS sub-cell capacity for the case that the whole AP capacity of the 2-Hop-Cell is made available only to one single FRS with varying FRS receive antenna gain. "Capacity" denotes the achievable aggregate cell throughput under the assumption of uniformly distributed MTs generating a constant bitrate type load[2]. The capacity of the AP (this case is equivalent to the AP operated as a conventional BS) amounts to 22.51 Mbits. The capacity that can be made available at the FRS, i.e., when the whole capacity of the AP is transferred to the area that is covered by one of the FRSs, amounts, depending on the FRS receive antenna gain, to values between 2.7 Mbits for 0 dBi gain and 15.87 Mbits for 30 dBi gain. The gap between the two curves in Figure 1 (right) denotes the capacity that has to be invested into the extension of the coverage range by means of relaying.

1.3   Realisation of MAC frame based Relaying - Example: HIPERLAN2

The HIPERLAN2 (H2) system is used here as an example to explain how MAC frame based protocols as 802.11e, 802.16a (HIPERMAN) and the recently adopted 802.15.3 can be applied to realise relaying in the time domain. All the MAC and PHY functions addressed here are existent in all these wireless standards and no changes of the existent specifications are needed for relaying. However, either the Logical Link Control (LLC) or MAC layer now needs a store-and-forward function like that known from a bridge to connect LANs to each other. In the description of a H2 relay we also use the term Forwarding when referring to Relaying. H2 specifies a periodic MAC frame structure, Figure 2. In the Forwarding Mode (FM) both signaling and user data are being forwarded by the FRS. An FRS operating in FM appears like a directly served MT to the AP. **Therefore, this does not preclude the possibility of allowing any MT to act as relay to become a Mobile Relay Station (MRS).** MTs are referred to as Remote MTs ((R)MTs) if they are served by a FRS.

The capacity of the MAC frame (see Figure 2, upper part) is assigned dynamically in a two-stage process[3].

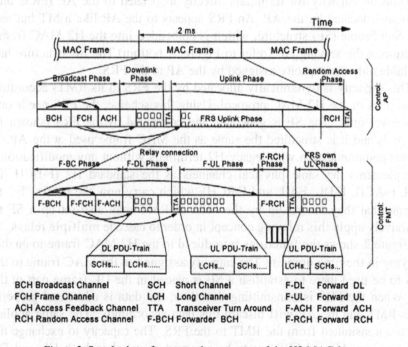

BCH Broadcast Channel   SCH   Short Channel   F-DL   Forward DL
FCH Frame Channel   LCH   Long Channel   F-UL   Forward UL
ACH Access Feedback Channel   TTA   Transceiver Turn Around F-ACH Forward ACH
RCH Random Access Channel   F-BCH Forwarder BCH   F-RCH Forward RCH

Figure 2. Standard-conformant enhancements of the H2 MAC frame

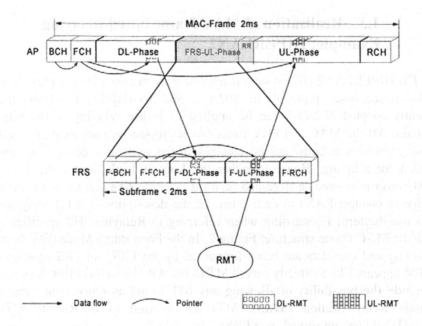

Figure 3. Data flow using a sub-frame in 2-Hop mode

Transmit capacity for terminals directly associated to the AP (FRSs and MTs) are allocated by the AP. An FRS appears to the AP like a MT but sets up a Sub Frame (SF) structure, which is embedded into the H2 MAC frame structure of the serving AP (refer to Figure 2, bottom).The SF structure has available only the capacity assigned by the AP to the FRS.

This capacity is dynamically allocated by the FRS to its RMTs according to the rules of the H2 MAC protocol. Using this scheme, the FRS needs one transceiver only. The SF is generated and controlled by the FRS (shown in Figure 3) and it is structured the same as the MAC frame used at the AP. It enables communication with legacy H2 terminals without any modifications. It implements the same physical channels as the standard H2 (F-BCH, F-FCH, F-ACH, F-DL, F-UL and F-RCH), which carry now the prefix "F-" to indicate that they are set up by the FRS. A RMT may also set up a SF to recursively apply this relaying concept in order to cascade multiple relays.

Figure 2 shows the functions introduced to the H2 MAC frame to enable relaying in the time domain. The capacity assigned in the MAC frame to the FRS to be used there to establish a SF is placed in the UL frame part of the AP. When the FRS is transmitting downlink, the data is addressed properly to its RMT and the AP will discard this data accordingly. The same applies for data transmitted from the RMT to the FRS. The capacity to exchange the data between AP and FRS has to be reserved as usual in both UL and DL directions on request by the FRS[3]. A very similar operation is possible by using the Hybrid Coordinator Access in IEEE802.11e[4].

2. ARQ-THROUGHPUT 1-HOP VS. 2-HOP

The question arises under what circumstances relaying would be beneficial, i.e. when a 2-hop communication is preferential to one hop. Figure 4 shows analytical results[5] comparing the throughput achieved with 1-hop and 2-hop transmission for the two scenarios depicted in the upper right corner of the figure under Line of Sight (LOS) radio propagation.

It is assumed that the FRS is placed at half the distance between the AP and the (R)MT. It turns out that from a distance of 370 m onwards, the 2-hop communication delivers a somewhat higher throughput than 1-hop, as marked by the shaded area.

Relay based 2-hop communication provides another considerable benefit already mentioned in Section 1.1: it is able to eliminate the shadowing caused by buildings and other obstacles that obstruct the radio path from an AP. An example of this is given by the scenario in Figure 4 together with the throughput gain (shaded) resulting from relaying.

In this scenario, the AP and the (R)MT are shadowed from each other by two walls that form a rectangular corner, e.g. a street corner. The COST259 propagation model (see Section 3.3) was used and the walls were assumed to have an attenuation of 11,8 dB each. The shaded area highlights that the 2-hop communication gains over one hop, starting at a distance of 30 m only.

The two examples establish that relaying is of advantage for both, increasing the throughput close to the cell border of an AP (under LOS conditions) and for bringing radio coverage (and throughput) to otherwise shadowed areas.

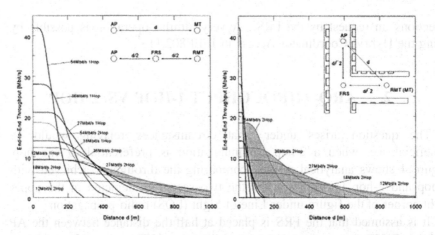

Figure 4. Comparison of the maximum achievable End-to-End Throughput over Distance for a 1- and 2-Hop Connection with ARQ

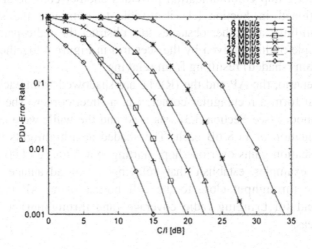

Figure 5. PDU-Error Probability for varying C/(I + N) and PHY-mode[8]

It has been explained by means of Figure 1 that relaying is consuming part of the capacity of an AP, since the relayed data has to go twice over the radio channel. It has been shown[6] that for relay based deployment concepts (like the one shown in Figure 6, left) MTs served at different relays that belong to the same AP can be served at the same time, whereby the capacity loss introduced by 2-hop communications can be compensated to a great extent. This capacity loss can even be turned into a substantial gain, if directive antennas are used at FRS as is shown in Section 4. Even if there is still a capacity loss resulting from a relay based system, this concept is able to trade the capacity available at an AP against range of radio coverage[7].

The trend towards increasing transmission rates resulting from further developed radio modems tends to provide an over capacity in the cell area served by an AP, especially in the first months/years after deploying a system. Relays substantially increase the size of the service area thereby increasing the probability that the capacity of an AP will be used effectively.

The next sections present a simulation-based performance evaluation of a relay-based system in a Manhattan-type environment.

3. SCENARIO AND SIMULATION ENVIRONMENT

3.1 Scenario: Dense Urban Hot Area Coverage

The dense urban environment with a high degree of shadowing has been identified as a scenario especially suited for deploying a relay based wireless broadband network. The Manhattan grid scenario[9] has been taken for the following investigations, see Figure 6. The most important parameters of the scenario are the block size of 200 m and the street width of 30 m. In the deployment scenarios without relays shown in Figure 7, each of the APs covers the range of two building blocks and one street crossing, resulting 430 m range.

This cell configuration requires a minimum of 4 carrier frequencies to ensure that in each direction, the co-channel cells are separated by at least one cell with another carrier frequency, see Figure 7 (left). Based on that structure, two possible variants can be considered. The APs can be placed at equal coordinates in adjacent streets, shadowed by the buildings (Figure 7, middle). The second variant is that APs are placed on street crossings (Figure 7, right), thereby covering horizontal and vertical streets. In this scenario without using relays, at least 8 frequencies are needed to ensure that co-channel cells are separated by cells using a different frequency.

All three scenarios shown in Figure 7 require that a cellular coverage in the Manhattan scenario would have to rely on LOS, leading to a high number of APs. Besides covering the scenario area with a single hop system, we study the impact of covering the same area with a system based on relaying. The basic building block, which consists of an AP and 4 FRSs is shown in Figure 6 (left). It has the potential to cover a much larger area than one Single-Hop AP. Figure 8 (top) shows the cellular deployment of these building blocks for various cluster sizes. Owing to the high attenuation caused by the buildings, only those co-channel interferers have to be taken into account that are marked in the figure in black and the reuse distance is indicated by the black arrows. For the cluster sizes N=2/3/4 we obtain reuse distances D=1380 m/2070 m/2760 m.

3.2 Scenario: Wide Area Coverage

The low coverage range that wireless broadband systems exhibit at high bitrates is shown in Figure 4 (left). In a conventional 1-hop hexagonal cellular approach, this leads to a large number of APs required for continuous coverage. It has already been suggested that the use of fixed relays can help to increase broadband radio coverage and thus reduce the number of APs needed. Figure 6 (right) shows the basic element (further referred to as "cell") used to achieve wide-area-coverage in a cellular approach. It consists of an AP and 3 surrounding FRSs which can be embedded into a hexagonal cell structure. We consider a coverage radius for a single AP or FRS of R=200 m. The result is that a relay based cell, which consists of 4 sub-cells has a radius of R=346 m. According to Figure 8 (bottom), different cluster sizes (N=3,7,12) can be realized just like in a traditional hexagonal cellular approach.

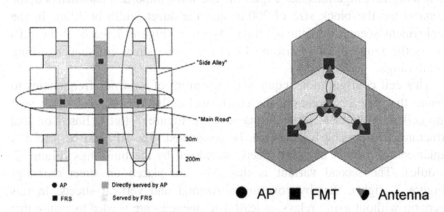

Figure 6. Left: Relay-based cell with four relays (below rooftop) in the Manhattan scenario
Right: Relay-based cell with three relays (above rooftop) in a wide-area scenario

3.3 Air Interface

All of the MAC frame based air interfaces mentioned above will operate in the 5 GHz licence-exempt bands (300 MHz in the US, 550 MHz in Europe, 100 MHz in Japan). We assume for the following studies that the physical layer (PHY) uses an OFDM based transmission with 20 MHz carrier bandwidth subdivided into orthogonal sub-carriers. The modem is assumed conformant to the IEEE802.11a standard. As indicated in Section 1, the 5 GHz frequency range is characterised by high attenuation and very low diffraction, leading to low radio range, which is one of the key problems addressed by the proposed relaying concept.

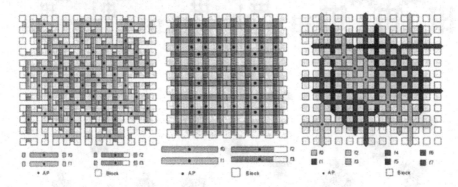

Figure 7. Possible AP deployments without relays

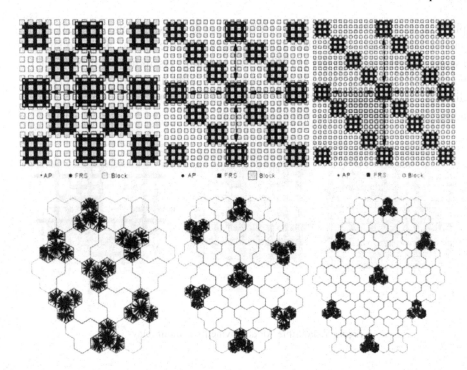

Figure 8. Top: City-wide coverage with relay-based cells for different cluster sizes N=2,3,4
Bottom: Wide-area coverage with relay-based cells for different cluster sizes N=3,7,12

Link-Level Performance: The basis for the calculation of the transmission errors is the ratio of Carrier to Interference and Noise power (C/(I+N)). The results of Link-level investigations[8] provide a Protocol Data Unit (PDU) error-probability related to the average C/(I+N) during reception of the PHY-PDU. This relation is shown in Figure 5. In our simulation model, collisions of interfering transmissions are detected and the resulting average C/(I+N) is calculated for each transmitted PHY-PDU to decide on success or retransmission.

Propagation Models: The COST259 Multi-Wall model has been used in the Manhattan scenario. This model[10] is an indoor propagation model at 5GHz, which takes into account the transmission through walls obstructing the LOS between transmitter and receiver. Unlike in the COST231 model[11], the attenuation non-linearly increases with the number of transmitted walls. Wall attenuations have been chosen according to the suggestions from the BRAIN project[12].

The Propagation Model used in the wide-area simulations is the Large-Open-Space model[9] and a pathloss exponent of $\gamma=2,5$ has been used.

Other Parameters of the Simulation Model: To determine whether a MT should be served by the AP directly or via a FRS, the path loss between

AP and MT is assessed. If it is higher than a certain threshold, the MT is associated to the closest available FRS ("closest" in terms of pathloss). The traffic load is assumed to be constant bitrate, which is a reasonable assumption when investigating the maximum achievable end-to-end throughput.

4. SIMULATION RESULTS

This section presents the performance evaluation results obtained by stochastic-event driven simulation. Results for the Downlink (DL) direction are presented here only, since the main effects that can be observed are quite similar in Uplink (UL) and DL directions, a result which is partly due to the Time Division Duplex air interface studied. We have also performed a mathematical analysis of the scenarios and the results validate the simulation results as visible from the results figures. This work will be published in the near future.

4.1 Reference Scenarios without Relays

In Figure 9 (left) the DL C/(I+N) and the related maximum End-to-End throughput are plotted over the distance of the MT from the AP when servicing the scenario by APs only, according to the Manhattan and the Wide-Area scenarios.

In the Manhattan Scenario, the C/(I+N) values are slightly higher for the deployment variant where the APs are placed on the street crossings (Figure 7, right) when compared to the other options. Figure 9 (upper right) shows the resulting Throughput (TP), again versus the distance of the MT from the AP. At distances of 115 m and 345 m, some additional interference on the crossings is visible for a deployment according to Figure 7 (right)

In the wide-area cellular deployment, the C/(I+N) values degrade as expected with decreasing cluster size. For comparison, also the C/(I+N) for a single AP without Interference is shown. Figure 9 (lower right) shows that at the cell border (at a distance of 200 m), a maximum End-to-End throughput of ca. 8 Mbit/s can be provided in the very optimistic case of N=19.

4.2 Simulation Results with Fixed Relay Stations: Manhattan Scenario

Simulations with fixed relays as introduced in Section 3.1 have been performed for the cluster sizes N=2/3/4, cf. Figure 8. Figure 10 (left) shows two sets of curves in one graph:

The C/(I+N) versus the distance of a MT from the AP (marked with 1. hop) and the C/(I+N) encountered by MTs being served by a FRS (marked with 2. hop). The FRS is located at a distance of 230 m from the AP on the "Main Road" (cf. the pictogram in the figure and Figure 6). This explains the peak of the C/(I+N) curve visible at that distance. Each set of curves has the cluster size N as a parameter. As expected, the curves with N=2 show the lowest C/(I+N) values. Figure 10 (right) shows the C/(I+N) situation in the "Side Alley" of the relay based cell. Like on the first hop, the situation for the MTs is almost similar to that of the MTs served directly by the AP in the single hop case, with the difference that the next LOS co-channel interferer is more than 780 m away, leading to lower interference and thus to a C/(I+N) which is approx. 4 dB higher than in the single hop case.

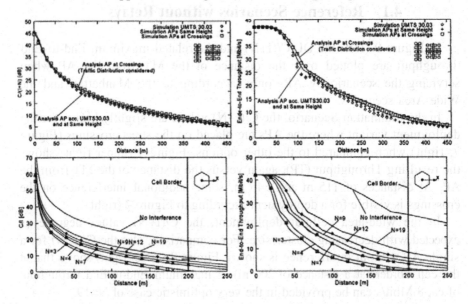

Figure 9. Top: C/(I+N) (left) and End-to-End-Throughput (right) without relays in Manhattan Scenario (Lines: analysis, Markers: simulation); Bottom: C/(I+N) (left) and End-to-End-Throughput (right) without relays in wide-area scenario (Lines: analysis, Markers: simulation)

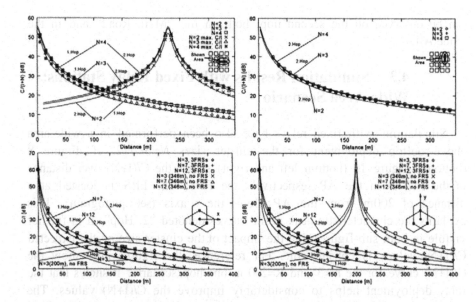

Figure 10. Top: DL C/(I+N) vs. Distance from (R)MT to AP respectively FRS for varying cluster sizes (2, 3, 4) using relays (Left: "Main Road", Right: "Side Alley", Lines: analysis, Markers: simulation); Bottom: Top: DL C/(I+N) vs. Distance from (R)MT to AP respectively FRS for varying cluster sizes (3,7,12) using relays (Left: "x-axis", Right: "y-axis", Lines: analysis, Markers: simulation)

A maximum TP of approx. 4-5,5 Mbit/s (depending on N) can be made available even at the cell border of the second hop, in an area which has no direct coverage of the first hop at all and which would require an additional AP in a single hop scenario (see Figure 7, right). Figure 10 (upper left and right) shows the resulting 2-hop TP for MTs on the "Main Road" and the "Side Alley" respectively with omnidirectional antennas used at AP, FRS and MTs. Obviously, the TP on both the first and the second hop depends on the cluster size N.

The relatively flat slope of the curves for the second hop indicates that the TP is upper-bounded by the capacity available at the FRS from the AP. More capacity can be provided when using gain antennas at FRSs and omni antennas at AP and MT, with the FRS serving its MTs with an omni antenna. The improvement in TP for the outer range of the relay based cell with an 11.8 dB gain at the FRS can be seen when comparing the left and right hand graphs in Figure 11. As predicted in Figure 1 (right), the resulting higher TP on the first hop allows a FRS to have much more capacity available in its service area.

At a gain of 11.8 dB, which is an intermediate value according to Figure 1 (right), an increase in max. TP of up to 80% (from 8 Mbit/s to 14 Mbit/s) can be observed on the second hop, both on the "Main Road" and in the "Side Alley".

4.3   Simulation Results with Fixed Relay Stations: Wide-Area Scenario

Simulations with fixed relays have also been performed in a wide-area above-rooftop deployment for the cluster sizes $N=3/7/12$, cf. Figure 8 (bottom). Figure 10 (bottom left and right) shows the C/(I+N) over distance of the MT from the AP respectively the FRS. The FRS is located at a distance of 200m from the AP along the y-axis (see pictogram). This explains the characteristic peak of the curves denoted "2. Hop". It is further visible in both sub-figures that the impact of the cluster-size on the expected C/(I+N) values is considerable. For reference, the figures also show the C/(I+N) curve for the N=3 and R=200 m one-hop scenario. It shows that the relay deployment helps to considerably improve the C/(I+N) values. The left-hand side of Figure 12 shows the maximum achievable Downlink End-to-End throughput versus the distance (in x- and y-direction) of a MT from the AP (marked with 1. Hop) and the throughput encountered by MTs being served by a FRS (marked with 2. Hop).

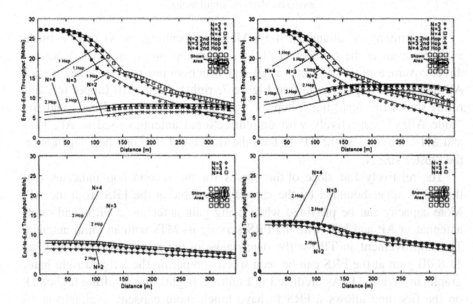

Figure 11. DL End-to-End-Throughput vs. Distance from (R)MT to AP respectively FRS for varying cluster sizes (2, 3, 4) using relays (left: using omni antennas only, right: with 11,8 dB receive antenna gain at the FRS (Lines: analysis, Markers: simulation)

The FRS are located at a distance of 200 m from the AP, e.g. in the y-direction (shown in the pictogram). This explains the maximum of the throughput curve for the second hop visible at that distance. Each set of curves has the cluster size *N* as a parameter. As expected, the curves with N=3 show the lowest throughput values, owing to the highest encountered interference. The right-hand side of Figure 12 shows the maximum achievable Downlink End-to-End throughput when an antenna gain of 11.8 dB is assumed between AP and FRS. Again, the upper figure represents the situation along the x-axis, while the lower figure refers to the y-axis of the relay based cell (also refer to the small pictograms included).

Like on the first hop, the situation for the MTs is almost similar to that of the MTs served directly by the AP in the single hop case (included for reference with a cell size of R=346 m).Depending on the cluster-size, the maximum End-to-End throughput along the y-axis improves for ranges greater than 220 m (N=3), 280 m (N=7) and 320 m (N=12) when relay stations are used instead of a single hop deployment. Along the x-axis improvements can be observed for N=3 and N=7 (ranges > 250 m and 325 m). In Section 2 we additionally show the result for the case where no co-channel interferers are present. In that case, improvements of the maximum throughput can be observed for distances greater than 370 m.

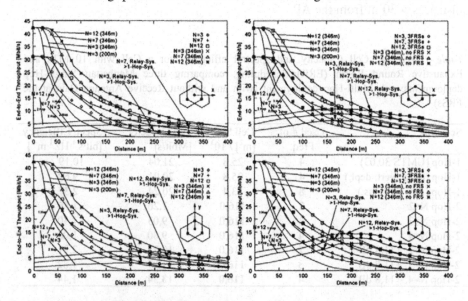

Figure 12. Maximum DL End-to-End-throughput vs. Distance from (R)MT to AP respectively FRS for varying cluster sizes (3, 7, 12) and sub-cell radii (200 m, 346 m) using relays (left: using omni antennas only, right: with 11,8 dB antenna gain between AP and FRS (Top: "x-axis", Bottom: "y-axis", Lines: analysis, Markers: simulation)

If an additional antenna gain is assumed between AP and FRS, the advantages of the FRS concept can already be observed at about 140 m (N=3), 170 m (N=7) and 190 m (N=12) along the y-axis, while - along the x-axis - the throughput of the two hop system outperforms the one-hop system starting at 170 m (N=3), 200 m (N=7) and 240 m (N=12). In general, a considerable improvement compared to the deployment without gain antennas can be observed. In addition, a more homogeneous distribution of the maximum achievable throughput can be noticed, which is especially beneficial in areas close to the cell border. The tighter the frequency reuse, the smaller becomes the minimal range where the use of FRSs is beneficial. Also, the number of necessary frequency channels is reduced with lower cluster sizes. This allows to use more frequency channels per cell and thus to increase an operators network capacity. When using FRSs, even in a cluster with N=3 the cell border can be served at sufficient quality due to the range extension. The gain obtained from the relaying scheme justifies transmitting the information twice.

The results given above are for the comparison of one- and two-hop cells with the same cell area (equal AP density). If an N=3-cluster with 200 m-cells is compared with a N=3 relay cell with 200 m sub-cells (equal site density), the advantages of the relay-based concept already become visible at distances > 30 m from the AP.

Table 1. Average Cell Capacity and spectral efficiency for a Cell with 10 MTs and Exhaustive Round Robin (ERR) Scheduling, comparing three Manhattan Single-Hop deployments with Multi-Hop deployment (with and without Receive Antenna Gain at the FRSs)

Scenario	Used # of Freq.	Cell Size $[m^2] / 10^3$	Cell Capacity [Mbit/s]	Spect. Efficiency $[bit \cdot s^{-1} \cdot Hz^{-1} \cdot m^{-2}]$
1-Hop (UMTS 30.03)	4	25,8	21,04	10,19
1-Hop (horiz./vert. depl.)	4	25,8	20,01	9,69
1-Hop (APs on cross.)	8	53,4	20,24	2,37
2-Hop N=2	2	116,0	7,26	1,56
2-Hop N=3	3	116,0	9,03	1,30
2-Hop N=4	4	116,0	9,80	1,06
2-Hop N=2, +11,8 dB	2	116,0	10,72	2,31
2-Hop N=3, +11,8 dB	3	116,0	12,7	1,82
2-Hop N=4, +11,8 dB	4	116,0	13,34	1,44

5. SYSTEM CAPACITY AND SPECTRAL EFFICIENCY

In addition to the End-to-End throughput studied in the previous sections, the system capacity, i.e. the aggregate traffic that can be carried in a well-defined service area and a certain amount of used spectrum is an important measure to assess a system's performance. To optimise a system, it is very important to have a clearly defined optimisation goal. The relay concept presented in this paper aims at providing a cost-efficient broadband coverage that can rapidly be deployed in a relatively large area. Table 1 shows the average End-to-End cell throughput for the different 1- and 2-hop deployments in the Manhattan scenario. The table also shows that the coverage area of one AP for the one-hop scenarios is relatively small, indicating that a large number of costly backbone connections is needed to cover the whole service area. From the small cell size and the high cell throughput results a relatively high area spectral efficiency. But a minimum of 4 carrier frequencies is needed in that case to provide continuous coverage. The AP deployment from Figure 7 (left) shows only small advantages over the horizontal/vertical placement (Figure 7, middle). The placement on street crossings (Figure 7, right) has the advantage that a larger area is covered per AP, reducing the number of needed backbone connections by a factor of 2. At the same time, a minimum of 8 carrier frequencies is needed to enable continuous coverage. This and the larger cell size lead to a substantial reduction in spectral efficiency, while the average cell throughput changes only slightly.

Table 2. Average Cell Capacity and spectral efficiency for a Cell with 10 MTs and Exhaustive Round Robin (ERR) Scheduling, comparing the wide-area cellular Single-Hop deployment with the Multi-Hop deployment (with and without Antenna Gain between AP and FRSs)

Scenario	Used # of Freq.	Cell Size $[m^2]/10^3$	Cell Capacity [Mbit/s]	Spect. Efficiency $[bit \cdot s^{-1} \cdot Hz^{-1} \cdot m^{-2}]$
Standard 200m	3	104	6,84	1,10
Standard 200m	7	104	12,2	0,84
Standard 200m	12	104	16,42	0,66
3FRS	3	311	4,21	0,23
3FRS	7	311	7,27	0,17
3FRS	12	311	9,46	0,13
Standard 346m	3	311	6,53	0,35
Standard 346m	7	311	11,42	0,26
Standard 346m	12	311	14,82	0,20
3FRS +11,8dB	3	311	7,44	0,40
3FRS +11,8dB	7	311	11,14	0,26
3FRS +11,8dB	12	311	13,41	0,18

Another reduction of the number of APs needed (to a total factor of 4) can be achieved by using FMTs as proposed in Figure 6. This leads to a very cost-efficient cellular coverage of the service area. The 2-hop transmission obviously reduces the cell capacity, an effect that can be reduced through the use of higher re-use distances.

A substantial increase in throughput and cell capacity is achieved through the use of directive receive antennas at the FMTs. When using 2 carrier frequencies, the relay concept with directive antennas achieves roughly the same area spectral efficiency $(2.31 \text{ bit} \times \text{s}^{-1} \times \text{Hz}^{-1} \times \text{m}^{-2})$ as the 1-hop deployment with APs on street crossings (2.37 bit×s-1×Hz-1×m-2), with the advantage of a lower number of APs and carrier frequencies needed.

Table 2 shows the average End-to-End cell throughput for the different 1- and 2-hop deployments in the wide-area scenario. Again, from the small cell size and the high cell throughput results a relatively high area spectral efficiency in the case of the 200 m-cells. However, the interesting observation is that the relay-based system achieves the same area spectral efficiency as a one-hop system with the same overall cell size. At the same time, as we have seen in Figure 12, the coverage quality at the cell border is superior in the two-hop case. Under dense frequency re-use (N=3), the two-hop system even exhibits a 14% higher spectral efficiency (compare lines 7 and 10 of Table 2).

6. CONCLUSIONS

Modern wireless broadband air interfaces are based on MAC frames, the only exemptions being IEEE802.11a/b/g but 802.11e uses a MAC frame, too. MAC framed air interfaces have been established in this paper to be useful for relaying in the time domain by just using the functions available from the existing standards. Deployment concepts using fixed relay stations have been shown to be of high benefit to substantially reduce the effort of interfacing APs to the fixed network (owing to a substantial reduction of APs needed). Relays have been proven to substantially extend the radio coverage of an AP, especially in highly obstructed service areas. Gain antennas at FRSs have been established to substantially contribute to increase the throughput at cell areas far away from an AP.

REFERENCES

1. J. Habetha, R. Dutar, and J. Wiegert, "Performance Evaluation of HiperLAN/2 Multihop Adhoc Networks," in *Proc. European Wireless,* Florence, Italy, Feb 2002, pp. 25–31.
2. T. Irnich, D. Schultz, R. Pabst, and P. Wienert, "Capacity of a Relaying Infrastructure for Broadband Radio Coverage of Urban Areas," in *Proc. 10th WWRF Meeting,* New York, USA, Oct 2003, http://www.comnets.rwth-aachen.de/
3. N. Esseling, H. Vandra, and B. Walke, "A Forwarding Concept for HiperLAN/2," in *Proc. European Wireless 2000,* Dresden, Germany, Sep 2000, pp. 13–17.
4. S. Mangold, "Analysis of IEEE802.11e and Application of Game Models for Support of Quality-of-Service in Coexisting Wireless Networks", PhD Thesis, Aachen University (RWTH), Aachen, 2003, http://www.comnets.rwth-aachen.de/.
5. N. Esseling, E. Weiss, A. Kraemling, and W. Zirwas, "A Multi Hop Concept for HiperLAN/2: Capacity and Interference," in *Proc. European Wireless 2002,* vol. 1, Florence, Italy, Feb 2002, pp. 1–7.
6. D. Schultz, B. Walke, R. Pabst, and T. Irnich, "Fixed and Planned Relay Based Radio Network Deployment Concepts," in *Proc. 10th WWRF Meeting,* New York, USA, Oct 2003, http://www.comnets.rwth-aachen.de/.
7. W. Mohr, R. Lueder, and K.-H. Moehrmann, "Data Rate Estimates, Range Calculations and Spectrum Demand for New Elements of Systems Beyond IMT-2000," in *Proc. of WPMC'02,* Honolulu, Hawaii, USA, Oct 2002.
8. J. Khun-Jush, P. Schramm, U. Wachsmann, and F. Wenger, "Structure and Performance of the HiperLAN/2 Physical Layer," in *Proc. of the VTC Fall-1999,* Amsterdam, The Netherlands, Sep 1999, pp. 2667–2671.
9. 3GPP, "Selection Proc. for the Choice of Radio Transm. Techn. of the UMTS, (UMTS 30.03)," ETSI, Sophia Antipolis, France, Report TR 101 112, V3.2.0, Apr. 1998.
10. L. M. Correia (Editor), "Wireless Flexible Personalised Communications, COST 259: European Co-operation in Mobile Radio Research", *COST Secretariat, European Commission,* Brussels, Belgium, Mar 2001.
11. E. Damosso and L. M. C. (Editors), "COST 231 Final Report - Digital Mobile Radio: Evolution Towards Future Generation Systems", *COST Secretariat, European Commission,* Brussels, Belgium, Aug 1999.
12. BRAIN, "D 3.1: Technical Requirements and Identification of Necessary Enhancements for HIPERLAN Type 2 Air Interface", *IST-1999-10050 BRAIN WP3 - Deliverable,* Sept. 2000.

Chapter 6

DYNAMIC ENHANCEMENT AND OPTIMAL UTILIZATION OF CDMA NETWORKS

JOSEPH SHAPIRA

Comm&Sens Ltd., 23 Sweden St., Haifa, Israel. Email: jshapira@comm-and-sens.com

Abstract: The basic cellular architecture for maximizing independent mobile multiple accesses was conceived over 50 years ago. The cellular architecture is revisited as applied to the CDMA air interface and in view of changes in the paradigm of the service from coverage-limited mobile traffic to a mix of urban pedestrian and in-building usage and from voice-only to a mix of data services. The salient features of CDMA, different than those of narrow band communications, provide opportunities for enhancement and optimization of the service by modifying the air access topology and control. These are reviewed, along with relevant experience.

1. INTRODUCTION

The cellular concept, incorporating a basically hexagonal cellular topology, each encompassing a hub (Base Transmission System – BTS), which may be further sectorized to three or maximally six angular sectors has been conceived for analog, FDMA communications over a relatively homogeneous environment. The dramatic progress in telecommunications technology and the change in paradigm from mainly car-mounted service to predominantly pedestrian and indoor service were not matched by respective developments in the cellular air interface topology.

The CDMA air interface has provided improvements with regard to reuse, capacity and coverage control, along with more localized dependency on the environment and traffic densities. CDMA offers new opportunities for enhancing the resource usage and performance by adding flexibility to the RF access topology, and tuning various system parameters in response to the state of the network. Some of the experience in applying enhancement and optimization techniques, accumulated during the 9 years of commercial CDMA service is reviewed, and lessons learned are discussed.

The first part of the paper reviews the CDMA air interface and interaction with the propagation channel, as a basis for the application of improvement techniques. Next, a review of enhancement techniques is provided. These techniques include ways of improving channel quality, reducing excessive Soft HandOff (SHO) regions and overhead, reducing excessive variations in the channel and link-loss and mitigating "radio holes", load balancing, and for dynamic resource control and optimization of the network.

2.   REVIEW OF THE CDMA AIR INTERFACE

The most distinctive feature of the CDMA cellular system is *the random code filtering of the reverse link (RL),* allowing for contiguous reuse of the same frequency over the network ("reuse of one"). The interference interaction is localized, opening flexibility for cells, sectors and auxiliary access points (e.g., repeaters) to match local densities and environments and yet not be constrained by the hexagonal, three-sectored topology. The "circle of influence" of a cell or a sector is limited to a little more then one ring of cells around it and tuning and optimizing of the network is localized.

The bandwidth of the system (1.25 MHz or 4.5 MHz for the different standards) provides partial frequency diversity and smoothing of the channel. The propagation channel conditions, predominantly flat Rayleigh in narrow band systems, are environmentally dependent in this system. The E_b/N_t required to meet the service objectives may vary from about 2 dB in a benign "Gaussian" channel to 11 dB with flat Rayleigh fading, within a single neighborhood, which affects all the major system performance parameters.

The flexibility of code protection level vs. data rate and latency opens yet a new field of resource management and optimization, with implications on the air interface.

The CDMA forward (FWL) and RL are different and react differently to the environment and load. They are not automatically balanced – a chore left for the tuning/ optimization process.

The capacity of the RL is interference limited, but independent on the location of the MSs (mobile stations) within the coverage area. The coverage depends on the MS available power, and out-of-cell interference. Both capacity and coverage of the FWL, on the other hand, depend on the total BTS available power, and *do depend* on the MSs' distribution within the controlled area.

The RL power control (RPC) controls the aggregate interference to the BTS, maximizes the RL capacity and minimizes the interference to other cells. The FWL power control (FPC) supports MSs suffering from deficient links. It may consume BTS power and add out-of-cell, and non-orthogonal in-cell, interference, and its dynamic range (range of the "digital gain") has to be limited.

The boundary of the cell on the RL, as defined by MS transmission that satisfies both cells, shrinks as the cell is loaded. The FWL boundary, where an MS measures equal Eb/Nt from both cells, expands for the cell that is loaded relatively more. The higher aggregate transmission power of the loaded cell adds excess interference (Nt) to the other cell (and non-orthogonal interference in-cell). The measure for entry (T_{ADD}) and departure (T_{DROP}) from the SHO zone is Ec/Io, relating only to the level of pilot transmission and the link loss. The reverse and FWL boundary is thus "spread out" based upon changes in the relative loads of the cells. The practice of maintaining the service quality by broadening of the "SHO window" (setting of T_{ADD}, T_{DROP}) increases the overhead and reduces the capacity.

3. CDMA AIR INTERFACE DESIGN AND OPTIMIZATION - CONSIDERATIONS

Channel improvement. The high values of required Eb/Nt due to flat fading conditions are mitigated by fast power control, which serves as a time-varying "matched filter" for the flat fading affecting slow moving subscribers. The deinterleaver decoder provides time diversity, which is effective for fast moving subscribers. These two are built into the system (IS-95 applies fast PC only on the RL). Additional types of diversity may be implemented on the forward link, and for remote RF access and repeaters, on both RL and FWL.

Reduction of excess SHO overhead. The SHO feature provides a smooth transition between cells, and links the RL Power Control of adjacent cells to an automatically controlled cluster of cells. It also provides macro and micro diversity, which enhance the link at the cell edge. However, FWL capacity is impaired by the multiple transmissions of the BTSs participating in the SHO process. Excess hardware resources are committed on the RL, similarly limiting the capacity. Control of the SHO area, in response to the state of the network, is therefore a necessary means for maximizing the network capacity.

Reduction of link loss variations within the coverage area. The complexity of the urban environment, including in-building service areas, creates localized areas with high link loss, prone to call drops and unsuccessful attempts. This loads the FWL and reduces its capacity, and stresses the outer loop of the RPC, reducing the RL capacity. Remote RF access points, repeaters and directional beams are considered for mitigation of these variations.

Load balancing between sectors and cells. Control of sector width and sector orientation alleviates the uneven load among sectors and maximizes the capacity of the cell. Beam tilt control affects the inter-cell load balancing.

Dynamic resource control and optimization. Modern networks undergo high variations in activity and balance during the day. A shift from mainly street-level activity during the day to mainly indoor activity in the evenings provides major challenges for network operators. The advent of data services further increases the rate of these changes. Dynamic control of the means of network control reviewed above, based on statistical network probes, is a viable means for continuously optimizing the network.

4. APPLICATION OF ADD-ON TRANSMIT DIVERSITY – REVIEW

Transmit diversity (TD) is achieved by transmitting the same signal from two antennas that are either spaced a de-correlating distance apart, or polarized orthogonally to each other. The burden of optimally combining the two signals lies with the MS receiver. The two transmissions are distinguished by different codes in the 3G standards. TD is not available in IS-95 standard). The application of TD as an RF add-on to operating BTSs, offering compatibility with both IS-95 and CDMA2000, is enabled by introducing an optimal time delay between the transmissions (Time Delay TD - TDTD), which allows the "rake receiver" at the MS to detect them separately and then optimally combine them. Alternatively, phase modulation of one of the branches (Phase Modulation TD - PMTD) generates a compound signal at the MS that is similar to selective diversity. Both techniques, in space or polarization diversity configurations, were successfully employed in different networks.

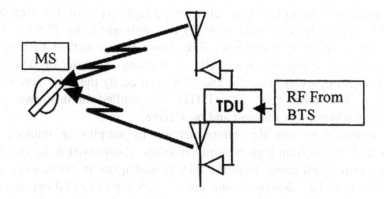

Figure 1 Transmit Diversity add-on

The basic architecture is described in Fig. 1. The transmit diversity unit (TDU) incorporates a power splitter, and a Time Delay or Phase Modulation unit, with respective low power amplifier to balance the branches.

The increase of the probability of dropped calls is a measure of the FWL reaching its power limit. Fig. 2 depicts a capacity increase of 68% with TD in a sector covering a US market residential area with 2 and 3 story buildings and trees nearby. TDTD was applied by splitting the RF transmit path and inserting a proper delay (in the TDU). The power of the main PA was reduced 3 dB and an additional equal power PA was added to the delayed branch, so as to maintain the same total power for comparison with the baseline. A space diversity antenna pair was used for both transmission and reception.

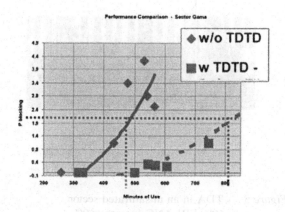

Figure 2 FWL Capacity increase in a saturated sector, due to transmit diversity.

An adjacent sector in the same cell covers a highway, with low link loss. The FWL capacity increase there with TD was measured to be 17.5%.

The measurement of average per-link power is one method to measure the capacity of a non-saturated cell. A capacity increase of 60% was obtained in Mexico City (Fig. 3a) and then verified by the same percentage increase in ERLANGs (Fig. 3b). TDTD was applied in this case, with polarization diversity on transmit and on receive.

An approach to provide indoor service by employing outdoor RF illumination suffers from high penetration losses. Compensation for this loss by excessive transmit power from the BTS is inadequate in urban areas with high penetration loss, leaving "radio holes" with high rate of dropped calls, while substantially increasing inter-cell interference and pilot pollution – all at a high cost of powerful power amplifiers. The benefit of an outdoor repeater, transmitting TD and providing RL diversity as well, positioned to

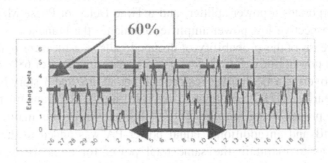

a

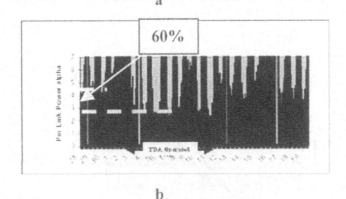

b

Figure 3   TDA in an unsaturated sector
    a.   60% ERLANG increase 60%
    b.   power decrease per-link

illuminate the building area, is shown in Fig. 4. An advantage of about 5 dB in FWL quality was repeatedly measured in walk tests indoors. This large improvement with TD results presumably from the low mobility and flat fading indoors. Respective reductions in the MS transmit power due to RL diversity is shown in Fig. 8.

Transmit diversity is thus shown to be an effective means for improving the FWL. It is most effective for slow movers in flat fading and high link loss areas. It increases both capacity and coverage of the FWL, and may need coverage adjustment to avoid cell extension, adoption of new users at the margin and a potential result of increasing rate of dropped calls.

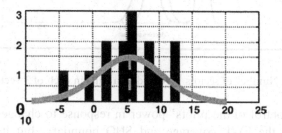

Figure 4   Histogram of difference (in dB) of received dominant pilot power (Ec) with application of Transmit Diversity

5.   CONTROL OF THE SOFT HANDOFF ZONE

The SHO zone is determined by the measurement of the respective pilots' power (T_{ADD} and T_{DROP}, respectively). These do not correlate with changes in the cell boundary on either the FWL or the RL due to changes in the load and relative loads between adjacent cells/ sectors. The practice of employing a broad SHO zone to absorb the changes is therefore common in CDMA networks, at the expense of system capacity. Fig. 5 depicts a representative histogram of the average number of sectors (pilots) involved in a call in each of 3 sectors in a cell, in a US market. One sector stands apart with 1.5 sectors involved, while the other two present typical values of 2.1 and 2.3. The respective power efficiency and capacity are 30% higher in the sector with lowest overhead.

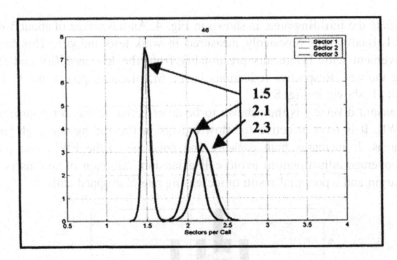

Figure 5   Number of sectors involved in a call, in each of 3 sectors of a cell.

Applying control of the pilots' power in response to changes in the link balance affects the FWL coverage and SHO boundary, but it also affects high link-loss areas within the coverage and increase "radio holes" and dropped calls. Beam tilting, on the other hand, controls the coverage, and steepens the link-loss slope, thus reducing the SHO area. This is illustrated in Fig. 6, where the distance from the BTS antenna is plotted on a LOG scale, and the free-space propagation is subtracted for presentation clarity. Power reduction (dotted) does not change the slope or the breakpoint, and weakens the link within the coverage area. Beam tilt, on the other hand, strengthen the link within the coverage area, steepens the slope and shrinks the distance to the break point. High sensitivity of the optimization to beam tilting has been shown repeatedly in different markets. Remote Electrical Tilt (RET) antennas ease the optimization process and enable the fine-tuning required.

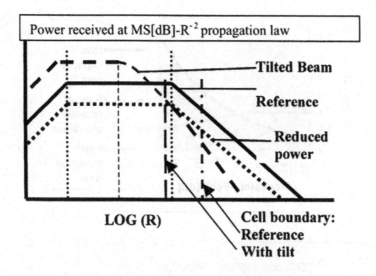

Figure 6 Comparison of the link-loss with power reduction and with beam tilt

6. REPEATERS IN CDMA SYSTEMS

Distribution of RF access points alleviates excessive variations in the link and appearance of "radio holes". The choice of means of transport of the signal to the access point depends on the cost of the various solutions: same channel RF transmission (f1/f1); other (unused) cellular frequency transmission (f1/f2) or microwave link; fiber optics or free-space optics. Full dedication of the cell/sector resources to the remote access point ("BTS hoteling") has only infrastructure deployment implications. For the air interface, it merely shifts the cell location to the access point. However, repeaters that share the sector resources at the RF transport level create interactive access points that affect the network control parameters.

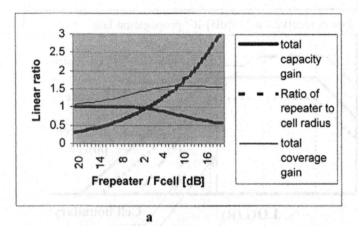

a

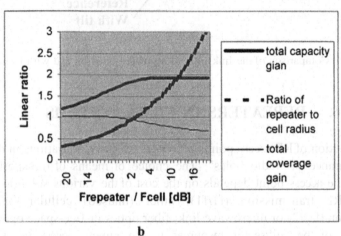

b

Figure 7 Repeater optimization. Repeater NF=Cell NF
    a.   Coverage-extension repeater. Cell load factor .7, repeater user density=1/3 of cell user density.
    b.   Capacity extension repeater. Cell load factor .4, repeater user density= 3 times cell user density

The control of the repeater gain (RL) provides a balance between the repeater and the donor cell coverage areas: The repeater gain determines the maximal link loss and repeater coverage. However the repeater gain also affects the noise rise at the donor cell due to the amplified repeater noise. Fig. 7 exemplifies the maximization of capacity and of coverage for different scenarios. The optimization of a "hot spot" repeater, embedded within the donor cell, depends on the user density distribution. The net gain for the coverage balance (where the direct link and the repeater link are equal), Net gain equal to the square of the ratio of the coverage areas, may be appreciably lower than unity, and is preferably set to exceed the high density ("hot spot") area.

Both transmit and receive diversity enhance the repeater channels and contribute coverage and capacity. An average improvement of about 4 dB is shown in Fig. 8 in the RL of indoor coverage area through a repeater illuminating the building area from the outside, due to Receive Diversity (RD). The usage of both TD and RD in the repeater provides a balanced enhancement, as shown in Fig. 4 and Fig. 8.

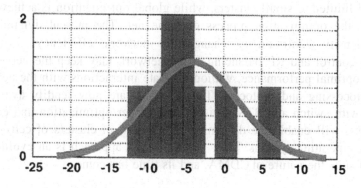

Figure 8   Reduction of MS transmit power (in dB) due to Receive Diversity in the repeater.

Diversity improvement is also provided between the repeater and the cell paths in the area of overlap between the repeater and donor cell coverage. However, the additional interference due to excessive multipath in this area may exceed the diversity gain benefit in some environments, and control of the overlap area is desirable.

Repeaters can be cascaded to cover a stretch of road. A chain of 3 repeaters may extend the distance between adjacent BTSs over 2.5 times, leaving over 35% capacity in the donor cell. A viable consideration for this application of RF repeaters is the limit on the repeater gain set by the isolation between the donor and distribution antennas. This is alleviated by using an unused RF channel for the transport (f1/f2), or electronic cancellation techniques.

7. DYNAMIC RESOURCE CONTROL AND OPTIMIZATION

The CDMA network achieves its capacity and performance through continuously averaging the interference, and allocating the power resources. It is automatically allocated throughout the network by the power control, linked by the SHO. The local nature of the interaction in CDMA limits the cluster of neighboring cells affected to a single ring around the change in the environment or in traffic densities. Optimization process is therefore also local and limited to small clusters, while global optimization is achieved by "sliding" the optimization process over the net. The present dynamics of traffic density continuously change the network efficiency. Dynamic coverage control and RF balancing of the network may keep the network at (almost) optimal performance, without delicate interactions with the network signal processing and protocols. The tools that are now readily available, together with switch and BTS statistics and proper optimization and control algorithms, will enable the dynamic optimization of clusters of cells as an add-on to operating systems. This "smart cluster" approach is an evolutional path that suits the nature of CDMA, and its time has come.

8. CONCLUSIONS

The CDMA network has been reviewed, with a focus on its interaction with the propagation channel. Means of enhancing the links, increasing coverage and capacity and controlling the network balance were reviewed. RF network control, incorporating controlled antennas, diversities and repeaters, and based on dynamic network statistics, will maximize the network utilization at all time. These techniques are available and suitable for add-on upgrading of the networks.

ACKNOWLEDGEMENT

The lessons reported herein have been learned through experience shared with operators and colleagues over the years. The author is indebted to Celletra Ltd. and its engineering team, with whom many of the solutions were developed and implemented, and to operators and their engineers that shared the experience. Standing out is Unefon AS, a Mexican operator, whose engineers ventured to successfully implement all these solutions.

PART II

WIRELESS LANS AND ADHOC NETWORKS

Chapter 7

ARCHITECTURE AND PROTOTYPING OF AN 802.11 BASED SELF-ORGANIZING HIERARCHICAL ADHOC WIRELESS NETWORK (SOHAN)

S. GANU, L. RAJU, B. ANEPU, S. ZHAO, I. SESKAR, D. RAYCHAUDHURI
WINLAB, Rutgers University, 73 Brett Road, Piscataway, NJ0885, USA

Abstract: This paper describes the design and implementation of a novel 802.11-based self-organizing hierarchical adhoc wireless network (SOHAN), and presents some initial experimental results obtained from a proof-of-concept prototype. The proposed network has a three-tier hierarchy consisting of low-power *mobile nodes* (MNs) at the lowest layer, *forwarding nodes* (FNs) with higher power and multi-hop routing capability at the middle layer, and wired *access points* (APs) without power constraints at the highest layer. Specifics of new protocols used for bootstrapping, node discovery and multi-hop routing are presented, and overall operation of the complete hierarchical adhoc network is explained. A prototype implementation of the SOHAN network is outlined in terms of major hardware and software components, and initial experimental results are given.

1. INTRODUCTION

This paper describes a novel self-organizing hierarchical adhoc wireless network ("SOHAN") designed to provide significant improvements in system capacity and performance relative to conventional "flat" adhoc networking approaches. The proposed hierarchical adhoc network is motivated by the fact that flat adhoc architectures do not scale well as the number of radio nodes becomes large (Gupta and Kumar, 2000). In addition, most realistic usage scenarios involve predominant mobile device traffic flows to and from the wired Internet, thus requiring effective integration of wired "access points" with the adhoc wireless network nodes.

The approach adopted here is based on a multi-tier hierarchy that scales well and integrates naturally with existing wireless access points or base stations, while retaining much of the robustness, coverage and power advantages of adhoc wireless networks. This architecture is applicable to a number of emerging adhoc networking scenarios including extended wireless local-area networks, home wireless networks and large-scale sensor networks. In each of these scenarios, the introduction of one or more tiers of adhoc forwarding nodes (FN) as intermediate radio relays between the MNs and APs helps to scale network throughput, reduce delay and lower power consumption at end-user devices. In this paper, we focus on practical design aspects and prototype implementation of protocols used in the SOHAN adhoc network, including those used for node bootstrapping, node discovery and multi-hop routing. More detailed consideration of system capacity scaling and network performance of the hierarchical adhoc network for alternative routing methods can be found in [Broch et al., 1999; Belding-Royer et al., 2002; Zhao et al., 2004). Design trade-offs for both discovery and routing are discussed, and a proof-of-concept prototype (implemented on Linux platforms with 802.11b radios) is described in terms of hardware and software components. Selected validation experiments and measurements are also given for the prototype hierarchical adhoc network.

2. SYSTEM ARCHITECTURE

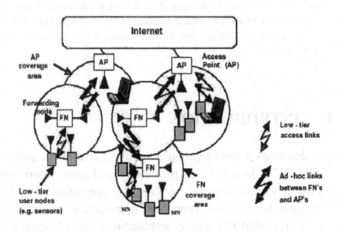

Figure 1 Three tier adhoc network architecture

The three-tier hierarchical adhoc network (Fig 1) consists of the following components: Low-power end-user "mobile nodes" (MN) at the lowest tier, higher powered radio "forwarding nodes" (FN) that support multi-hop routing at the second level, and wired access points (AP) at the third and highest level. Each of the network entities in the proposed system is defined in further detail below

- *Mobile Node* (MN) is a mobile end-user device (such as a sensor or a personal digital assistant) at the lowest tier (tier 1) of the network. The MN attaches itself to one or more nodes at the higher tiers of the network in order to obtain service using a discovery protocol. The MN uses a single 802.11b radio operating in adhoc mode to communicate with the point(s) of attachment. As an end-user node, the MN is not required to route multi-hop traffic from other nodes. It is noted that as a battery-operated end-user device, the MN will typically have energy constraints

- *Forwarding Node* (FN), is a fixed or mobile intermediate (tier 2) radio relay node capable of routing multi-hop traffic to and from all three tiers of the network's hierarchy. As an intermediate radio node without traffic of its own, the FN is only responsible for multi-hop routing of transit packets. A forwarding node with one 802.11 radio interface uses the same radio to connect in adhoc mode to MNs, other FNs and the higher-tier APs defined below. Optionally, an FN may have two radio cards, one for traffic between FNs and MNs and another for inter FN and FN-AP traffic flows (typically carried on a different frequency). The FN is typically a compact radio device that can be plugged into an electrical outlet, but in certain scenarios, may also be also be a battery-powered mobile device. Thus, the FN is also energy constrained, but the cost is typically an order of magnitude lower than that of the MN defined above

- *Access Point* (AP) is a fixed radio access node at the highest tier (tier 3) of the network, with both an 802.11 radio interface and a wired interface to the Internet. The AP is capable of connecting to any lower tier FN or AP within range but unlike typical 802.11 WLAN deployments, it operates in adhoc mode for each such radio link. The AP also participates in discovery and routing protocols used by the lower tier FNs and MNs, and is responsible for routing traffic within the adhoc network as well as to and from the Internet. Logically, the tier 3 APs are no different from tiers 1 and 2 when routing internal adhoc network traffic - the wired links between APs are reflected in (generally) lower path metrics. Since the AP is a wired node, it is usually associated with an electrical outlet and energy cost is thus considered negligible

3.   SCALABILITY OF HIERARCHICAL
NETWORKS

Scalability issues for flat adhoc networks as addressed by Gupta and Kumar (2000) motivate our proposed hierarchical architecture with more than one tier of adhoc radio nodes in which the lower tiers aggregate the traffic up to the intermediate relay nodes, while continuing to use robust adhoc self-organization and routing protocols. In order to study the performance of the hierarchical architecture under consideration, the performance of two routing protocols (DSR and AODV) when applied to a hierarchical and a conventional flat adhoc network was compared using ns-2 simulations. From Fig 2, it can be seen that the system capacity increases significantly when a hierarchical approach is adopted for the particular system example under consideration. Similar gains are observed for both DSR and AODV. Performance measures such as delay and packet delivery ratio are also improved in the hierarchical system. Further details can be found in (Zhao et al., 2004), in which the authors also study the scalability of the three-tier hierarchical network's capacity as a function of the relative densities of FNs and APs.

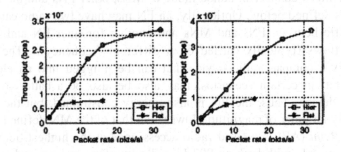

Figure 2   Performance of a) DSR and b) AODV applied to a flat and hierarchical network

Results indicate that it is possible to scale network capacity quite well with a mix of several (lower-cost) radio forwarding nodes and just a few wired access points.

4.   ADHOC NETWORK PROTOCOLS FOR
SOHAN

The above results motivated the design and development of a proof-of-concept prototype for the proposed self-organizing hierarchical adhoc network (SOHAN).   The adhoc protocols used in the hierarchical network including those meant for 1) Bootstrapping, 2) Discovery, 3) Routing and Data Transmission are described below.

4.1 Bootstrapping

This phase involves the configuration of the different devices in terms of channel assignments and initial transmit power level settings. Note that the devices operate in the 802.11 adhoc mode (IEEE 802.11 standards, 1999).

- Each AP is initialized on a pre-determined channel
- Each FN has two interfaces, one to communicate with other FNs and MNs (known as the *beaconing* interface) the other interface to communicate with APs (known as the *scanning* interface). These two interfaces are configured to operate on different channels that are specified at initialization so as to minimize interference.

In (Raju et al., 2004a), a distributed bootstrapping mechanism that will automatically select appropriate channels for the particular interface based on the number of nodes already existing on that channel has been proposed. However, for the current implementation, the channel allocations are done manually using scripts.

4.2 Discovery

In traditional adhoc networks, there is no discovery phase and the routing protocol itself is responsible for building up topologies either using on-demand broadcast of route requests or by exchanging neighbor information proactively with one hop neighbors. While this may be sufficient for smaller networks, as the number of nodes increases, it results in denser physical topologies, leading to extensive routing message exchanges. The problem is more severe in a multi-channel network where the multiple nodes that need to communicate could be on different radio channels. In this case, the routing messages need to be propagated across multiple channels in order to enable data transfer from one node to the other. In [Raju et al., 2004b], using ns-2 simulations, it has been shown that by introducing discovery as a separate layer, the routing overhead is significantly reduced. These results, as shown in Fig. 3, demonstrate the improvement in routing overhead versus varying mobility and number of nodes with AODV as the routing protocol.

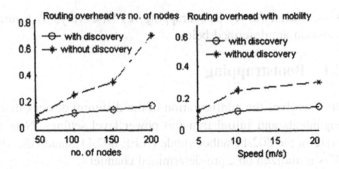

Figure 3 a) Routing overhead with increasing nodes
b) Routing overhead with increasing mobility

Based on these results, we use augmented 802.11 MAC beacons and associations in SOHAN to support neighbor discovery and determination of the logical topology. Note also that for ease of implementation, the beacons used in the prototype are application-level packets, since actual 802.11 beacons are generated by the firmware in most of the existing 802.11b network adapters and are not customizable. The beacon format in SOHAN is shown in Fig 4.

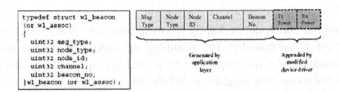

Figure 4 Beacon and Association Message Format

In SOHAN, FNs and APs periodically send beacons while the MNs scan different channels listen to the beacons and send an association message to the best "cost" parent using the discovery metric described below.

4.2.1 Discovery metric

For our implementation, energy conservation at the MNs was chosen as the objective and the discovery metric was based on minimizing the transmit power consumption at the MNs. We modified the device drivers to append transmit power to each outgoing beacon at the APs/FNs and the received signal strength for each incoming beacon at the MNs. Using this information and assuming reciprocity of channel, the node with the minimum transmit power was chosen as the next hop neighbor. In case, there were two or more such nodes, the node whose beacon was received with the higher signal strength was chosen.

4.3 Routing

Motivated by the results in Fig. 3, we have implemented a distance-vector based routing protocol that uses the "logical" topology information presented by the discovery mechanism in order to create and maintain local neighbor tables at each of the FNs and APs. A combination of MAC addresses and node ID of the nodes is used for the routing protocol to handle the case of FNs that have two different MAC addresses for the two different interfaces but the same node ID. The routing protocol involves two phases: 1) Neighbor Table Formation and 2) Periodic Table Update and Exchange. In phase 1, the FNs and APs build their local neighbor tables based on the beacons and the association messages exchanged during the discovery phase. The neighbor table format is shown in Table 1. Each entry is associated with a refresh timer that is reset or decremented respectively based on whether or not beacons are received from that neighbor every beacon interval.

Table 1. Local Neighbor Table Format

MAC Addr.	Node Type	Purge Timer	Channel to next hop	Cost to dest.	Interface to next hop	Next hop

During phase 2, FNs and APs exchange their local neighbor tables amongst themselves using sequence numbers to handle loops and update their neighbor tables based on this exchanged information. The MNs are not involved in the routing mechanism and simply forward their data to the best cost parent selected by the discovery procedure.

Note that any existing proactive (DSDV (Perkins and Bhagwat, 1999)) or reactive (DSR, AODV) routing mechanism can be implemented on top of our discovery mechanism. For DSDV, the "forwarding table" at each node could be replaced by the neighbor table provided by the discovery protocol; while in AODV (or DSR), the route requests could be propagated only to a subset of nodes as selected by the discovery mechanism.

After the table exchanges, each FN computes a path that it can use to route data to the AP. The FN maintains a latest best cost path towards the AP at every instant and whenever it receives data packets (originated at the sensors), it consults the neighbor table to forward the data to the next hop on the appropriate channel and interface. If an FN is disconnected from the network (there is no entry for an AP that exists in its neighbor table), it discards the packet and indicates a routing failure. Note that this routing implementation is based on the assumption that most of the traffic flow is from the MNs to the APs.

5. PROTOTYPE IMPLEMENTATION OF SOHAN

In this section, the practical design aspects and implementation of a proof-of-concept prototype for SOHAN architecture are described. The software architecture, protocol details, hardware architecture and initial experimental results are discussed in detail.

5.1 Software Architecture and Protocol Implementation

The implementation of the discovery and routing mechanism was done using C programming on embedded devices running Linux. We used the Libnet open-source library to generate, send and receive custom packets. Fig. 5 shows the software architecture of the prototype. The modular software design as described below is consistent with the protocol stack and thus provides an easy way to modify functionality and add features at any layer.

- **Physical Layer**: The functionality of transmitting and receiving packets is handled using Libnet packet handling library that provides a portable and simplified interface for low-level network packet shaping, handling and injection.

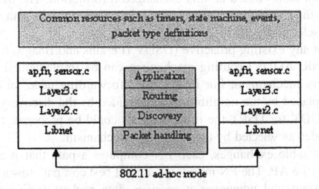

Figure 5 Software architecture of SOHAN

- **Layer2.c:** This layer handles the discovery and MAC layer functionality. Whenever a packet arrives from the lower layer, this layer handles the packet processing and passes the information to the higher layers.
- **Layer3.c:** This layer is responsible for handling the maintenance of the local neighbor tables and periodic exchange of neighbor tables amongst one-hop neighbors. The neighbor table is maintained and updated based on the beacons and the associations that are received from layer2.c. Upon the expiration of the route update timer, a periodic neighbor table exchange takes place. Entries are purged upon expiration of the refresh timer.
- **Application Layer** (Sensor.c, fwnode.c, ap.c): This layer handles the application specific functionality that depends on the type of the nodes.
- **Common functions:** The common functionality such as timer management, event management, finite state machine, packet type definitions and common wireless utilities is handled by programs common to all layers.

5.2 Hardware Platforms Used

The APs were based on a US Robotics 2450 Access Points running customized AP code for adhoc mode. The FNs were built on Compulab 586 CORE platform running a 133 MHz processor with two PCMCIA slots for two wireless interfaces. The MNs were built on the embedded Cerfcube platform that was battery operated and ran our custom sensor application. The selection of the hardware platforms was consistent with the system architecture and operated under the same set of constraints at each tier. Fig. 6 captures the different platforms used for the devices.

AP   FN   SN

Figure 6 Hardware platforms used for SOHAN

6. EXPERIMENTAL RESULTS

The experimental nodes ran Linux (kernel 2.4.17) with device driver modifications for recording and appending transmits power and received power to every outgoing and incoming packet respectively. We ran simple tests to determine appropriate values for parameters such as beaconing interval, and channel dwell time prior to conducting our benchmark experiments.

6.1 Discovery delays versus beacon interval and dwell times

The MNs were configured to scan every channel and varied the channel dwell times (from 100 ms to 1 sec) for different experimental runs. At the AP, the beacon interval was varied from 100 ms to 500 ms. As described in section III.B, the beacons were generated at the application layer and injected into the card using Libnet packet library. Also, scanning across channels at the MNs was performed at the application layer using *ioctl* calls to the device driver. We measured the discovery delays for a scenario consisting of a single AP and MN. This was repeated for different beacon intervals (100 ms, 250ms and 500 ms) at the AP with varying dwell times (from 100 ms to 1 sec) at the MNs.

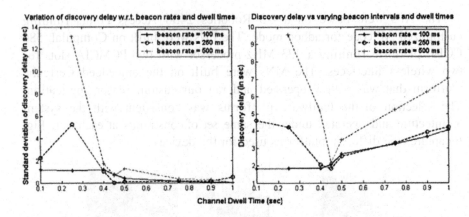

Figure 7 a) Discovery delays with different dwell times and beacon intervals
b) Variance of discovery delay

Discovery delay is the time interval between beginning the experiment (both nodes starting at the same time) until the AP received the first 'association' message from the MN. Figure 7a show the results for the average discovery delays (in sec) for several sample runs for each setting along with the standard deviation of the delays. As shown in Figure 7b, for dwell times below 450ms, the discovery delay showed a high variation. This was because the application at the sensor missed a lot of beacons during its scan and hence the time instant at which the first beacon was received was highly variable. When the dwell time per channel was higher than 450 ms, the variation of the discovery delay is significantly lesser than in the previous case. The performance with beacon intervals of 100ms and 250 ms was very similar. Hence, the beacon interval at the APs/FNs was chosen to be 250 ms with a channel dwell time of 450 ms at the MNs as a compromise between discovery delays and injecting more beacons in the network, which increased the discovery overhead.

6.2   Packet delivery ratio and average delays

In this experiment, the MNs transmitted at varying data rates to the AP over the hierarchical network. Two different packet sizes (1024 and 1472 bytes, UDP) were used. Fig. 8a shows that the packet delivery ratio for moderate loads is high. The small loss of packets may be attributed to the forwarding node rediscovery period, during which all packets received at the FN are dropped. For higher loads, the network degrades to deliver only 50 percent of the offered data. This is largely due to packets being dropped at the transmitter's interface. We also noticed that in such conditions, the application level beacons that we use were also dropped. This resulted in extremely large discovery times, which further amplified the problem of packet loss. However, using firmware-generated beacons should solve this problem. Fig. 8b shows that the end-to-end delay even for moderate loads is high (on the order of a second) which can be attributed to relatively high software latency with the embedded devices used. We note that this delay is the time elapsed between the MN application layer sending data and the AP application layer receiving data and includes system delays at the transmitter, receiver and switching between two interfaces at the intermediate FN. However, the system is tolerant to delays under increased traffic loads, which is due to the presence of two radio interfaces at each FN operating on different channels.

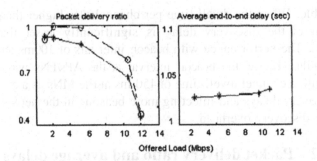

Figure 8 a) Packet delivery ratio
b) average delay of the network with increasing offered loads

A similar flat 802.11b adhoc network will tend to have a lower system capacity due to larger hop counts and a single frequency. These benchmark results indicate that the hierarchical network prototype we have developed provides promising results, which are fairly consistent with predictions from simulation. It is observed that implementing the discovery protocol in firmware will result in better network performance.

7.   CONCLUSIONS AND FUTURE WORK

We have presented the architecture and prototyping of SOHAN, a hierarchical self-organizing wireless adhoc network consisting of 802.11 based heterogeneous radio nodes at the three tiers. The architecture is motivated by potential improvements in scalability and system performance when compared with conventional flat adhoc networks. A proof-of-concept prototype was developed for evaluation of protocol design options and validation of system performance. Experimental results obtained so far are fairly consistent with predictions from simulation, and are indicative of the advantages of the proposed hierarchical structure with self-organizing discovery and routing protocols. Topics for future work include further optimization of discovery and routing algorithms, mobility support and joint MAC/routing methods for capacity and quality-of-service improvements.

REFERENCES

E.M. Belding-Royer, "Hierarchical Routing in Adhoc Mobile Networks", *Wireless Communication and Mobile Computing,* pp.515-532, 2002.

J.Broch, D.Maltz and D.Johnson, "Supporting Hierarchy and Heterogeneous Interfaces in Multihop Wireless Adhoc Networks", *Workshop on Mobile Computing,* June 1999

P. Gupta and P.R. Kumar, "The Capacity of Wireless Networks", *IEEE Transactions on Information Theory,* vol. 46, March 2000, pp. 388-404.

IEEE 802 LAN/MAN Standards Committee, "Wireless LAN medium access control (MAC) and physical layer (PHY) specifications", IEEE Standard 802.11, 1999

Libnet Packet Library, http://www.packetfactory.net/ libnet/

C. E. Perkins and P. Bhagwat, "Highly dynamic destination-sequenced distance-vector routing (DSDV) for mobile computers", *Proc. ACM SIGCOMM' 94,* pp 234-244.

L. Raju, S. Ganu, B. Anepu, I. Seskar and D. Raychaudhuri, "BOOST: A Bootstrapping Protocol for Self-Organizing Hierarchical Adhoc Networks", *IEEE Sarnoff Symposium,* April 2004a.

L. Raju, S. Ganu, B. Anepu, I. Seskar and D. Raychaudhuri, "Beacon Assisted Discovery Protocol for Self-Organizing Hierarchical Adhoc Networks", under review at *IEEE Globecom 04,* Nov. 2004b.

S. Zhao, I. Seskar and D. Raychaudhuri, "Performance and Scalability of Self-Organizing Hierarchical Adhoc Wireless Networks", *to appear in the Proceedings of the IEEE WCNC 2004,* March 21-24, 2004, Atlanta.

Chapter 8

VOICE CAPACITY IN IEEE 802.11 NETWORKS
Experimental Results and Analysis

MONCEF ELAOUD[1], PRATHIMA AGRAWAL[2]

[1]Internet Research Lab, Telcordia Technologies, One Telcordia Drive, Piscataway NJ 08854-4102, USA, moncef@research.telcordia.com; [2]Wireless Engineering Research and Education Center (WEREC), Department of Electrical and Computer Engineering, Auburn University200 Broun Hall, Auburn, AL 36849-5201,USA, pagrawal@eng.auburn.edu

Abstract: We are currently witnessing a boom in the deployment and usage of Wireless Local Area Networks (WLANs). Once only seen within the enterprise, WLANs are increasingly making their way into residential, commercial, industrial, and public areas. The recent efforts of carriers to integrate WLANs into their wide-area service offerings are testimony to their growing role in the future of wireless networking. Voice continues to be the most predominant wireless application and in order to fully integrate with existing and future cellular systems WLANs must deliver high quality voice service. In this chapter, we present experimental results of voice over IP (VoIP) capacity in an IEEE 802.11b. Results show that less than 6 voice conversations can be supported in an IEEE 802.11b. This small voice capacity represents a great hurdle to Wi-Fi service providers who want to integrate voice services into their public offerings. In this chapter, we identify and quantify the contributing factors that limit the voice capacity in IEEE 802.11 and provide some insights on how the capacity can be enhanced.

1. INTRODUCTION

Wireless Local area Networks (WLANs) facilitate easy collaboration and efficient communication on the fly without the need for costly network infrastructure. The most popular WLAN technology is the IEEE 802.11 standard also known as Wi-Fi. Once only seen within the enterprise, WLANs are increasingly making their way into residential, commercial, industrial, and public areas. Traditionally, these networks serve a floating end user population that demands flexibility in connectivity such as in hotels, airports, coffee shops, etc. A number of wireless service providers, such as Boingo Wireless, iPass Inc. and Sputnik Inc. are attempting to knit together a national patchwork of such local-area networks to provide connectivity. In addition, there are new efforts to deploy WLANs as supplement or replacement to existing wideband technologies such as DSL. Recently, the Los Angeles suburb of Cerritos, California, became the biggest 802.11 hot spot in the U.S (Evers 2003). AirNet Wireless LLC deployed an IEEE 802.11 network covering 8.6 square miles to serve some 50,000 residential, city, and commercial users who had no access to traditional wideband Internet service such as DSL and cable.

As a result of the fast proliferation of WLANs, there is an increasing thrust to integrate WLANs with other existing wireless technologies such as 3G. An initiative led by established wireless service providers including Verizon, AT&T Wireless, and T-Mobile to provide national WLAN service for business travelers seeks to capitalize on the benefits of integrating WLANs into their public service offerings. While a portion of the revenue of these systems is expected to result from wireless data, we expect that voice will remain a predominant application and be a significant driver for adoption and integration. Additionally voice applications are especially important in vertical industries such as construction, healthcare, and banking.

As a consequence, there have been several attempts to provide voice service over IP over 802.11b networks. Companies such as SpectraLink and Meru Networks have commercial solutions available. Despite the large available bandwidth in these networks (up to 11 Mbps) voice capacity remains small. Several studies have uncovered the limited voice capacity afforded by IEEE 802.11b. Elaoud and Anjum (2003) have shown through testbed experiments that a standard IEEE 802.11b access point can only support about 10 voice calls. SpectraLink (2004) also reports that the maximum number of supported voice calls by an IEEE 802.11b access point is less than 10 calls. Veeraraghavan (2000) and Koepsel (2001) show similar results through simulations.

The design of MAC protocols for supporting voice traffic has drawn some attention in the literature (Kubba 1997; Akyildiz 1999). Much of the previous work covering voice over IEEE 802.11 considers simulation of the PCF operating mode (Kubba 1997; Akyildiz 1999; Stine 1998; Chen 2002).

It is not until the last few years that experimental studies have made their way into the technical literature. Garg and Kappes (2003) compare experimental voice capacity performance to that predicted by analysis. Their experimental testbed comprised of commercial off the shelf equipment and used a voice codec rate of 10ms. Each voice conversation contained a wired and a wireless participant. The experiment revealed that 6 such conversations could be reliably supported, but that when the 7th call was added, performance for all flows in the wired-to-wireless direction suffered unacceptable losses and delays. Thus the authors concluded that capacity for a 10 ms codec is 6 terminals. However, the authors did not detail exactly what performance metrics constitute an "acceptable" voice call. In addition, they did not consider the impact of background data traffic on voice calls.

In contrast, Elaoud and Anjum (2003) describe a quantifiable metric, though it is based solely on packet loss and did not account for the effects of delay on voice performance. This constraint was due to experimental limitations in achieving time-synchronization among senders and receivers, without which one cannot accurately measure packet delays.

In the present work, we present testbed measurements that overcome these limitations and allow us to accurately measure one-way packet delays. This enables a more detailed measurement of voice capacities that incorporates both packet losses and delays. Using these methods, we derive a voice capacity baseline in 802.11b networks using standard, off-the-shelf equipment. In addition, we present simple analysis to understand the contributing factors to the limited voice capacity. Specifically, we evaluate the effects of packet header overhead, access overhead and the inherent access fairness of 802.11 on voice capacity.

The rest of the chapter is organized as follows. Section 2 we present the access mechanism of IEEE 802.11 standard. Section 3 defines the quality of service metric used in this chapter to evaluate the performance of VoIP sessions. We also define voice capacity in terms of packet loss and delay. Section 4 presents our testbed setup and highlights our experimental results. The analysis of the results is presented in Section 5 and Section 6 where we present effects of header, access overheads and fairness on voice capacity. The chapter concludes in Section .7

2. ACCESS MECHANISM IN STANDARD 802.11

The IEEE 802.11 provides two modes of access namely the required Distributed Coordination Function (DCF) and the optional Point Coordination Function PCF. The DCF access mode is based on the Carrier Sense Multiple Access with Collision Avoidance (CSMA/CA) family of medium access protocols. These MAC protocols are designed to avoid collisions through a "listen-before-talk" procedure. Terminals must first determine if the medium is free before attempting to transmit. To account for the vagaries of the wireless channel and the possibility of packet loss,

MAC frames must be positively acknowledged by receiving stations. Transmitting stations that do not receive such an acknowledgement consider the transmitted frame lost and retransmit.

Figure 1 depicts the state diagram used by an IEEE 802.11 station to transmit its packets. A station that has packets to transmit first senses the medium. If the medium is determined to be free for a continuous period equal to a Distributed Inter Frame Spacing (DIFS), the station transmits the packet. Otherwise the station continues to monitor the medium until that condition is met. At this time the station enters a random backoff phase in which it chooses a random backoff timer uniformly from a collection of values known as the contention window. If multiple stations want to transmit a packet, the station with the lowest random backoff timer wins the contention for the medium. IEEE 802.11b standard specifies the minimum contention window to be CWmin = 31 time slots and the maximum to be CWmax = 1023, where a time slot is defined to be 20 microseconds.

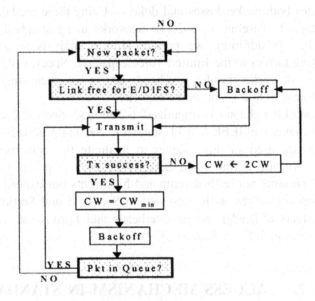

Figure 1   Access Mechanism in IEEE 802.11

The backoff procedure is shown in Figure 2. During the backoff procedure, the station continues to monitor the medium and for every idle timeslot decrements the backoff timer. If the medium becomes busy during the countdown, the station suspends the decrement operation until the channel becomes idle again for a period of DIFS. When the backoff timer reaches zero the station transmits its packet. After the completion of a packet transmission, successfully or unsuccessfully, the transmitting station enters a post-transmission random backoff procedure. After every unsuccessful packet transmission the size of the contention window of the failing station, and thus the average waiting time, doubles until it reaches its maximum value. However, following a successful transmission the contention window range is reset to its minimum value.

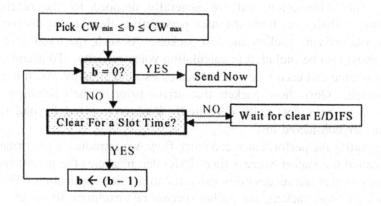

Figure 2 Backoff Procedure in IEEE 802.11

Upon the correct receipt of a packet, the receiver station sends a positive acknowledgement. Absence of a positive acknowledgement indicates lost or corrupted transmissions to the sender. Following a failed transmission the sender may retransmit the packet up to a maximum number of times before it is dropped. For more details on the DCF mode of 802.11b we refer the reader to ISO/IEC (1999) and IEEE STD 802.11b (1999).

3. VOICE QOS METRICS AND NETWORK CAPACITY

Deriving perceived voice quality based on objective traffic measurements is, in general, difficult. Some complex models exist, such as Perceptual Evaluation of Speech Quality (ITU-T P862 2001) and others based on the ITU E-Model (ITU-T G107), that attempt to account for transients and temporal correlations in channel performance. However, as a first approximation, we have chosen a simple, straightforward model based solely on per-call packet loss and per-call packet delay. Our data collection methodologies, however, do not preclude further refinement of the results via the above perceptual models and such a comparison is scheduled for future work.

Voice over IP is a time and loss critical application where a certain percentage of data packets must be delivered within a certain maximum tolerable delay. Standards bodies have suggested that a voice flow can sustain up to 2% packet loss and tolerate up to 200 milliseconds of one-way end-to-end delay and still deliver acceptable voice quality (ETSI TR 101). Packets that are delivered beyond the application specific deadline usually contain stale information and are generally dropped by the receiving application. Therefore, from the user perspective, there is no distinction between late arriving packets and lost packets. As such, the fraction of late packets must also be included in calculating voice capacity. To accomplish this, we assume that each voice packet possesses a deadline by which it must be delivered. Only those packets that arrive prior to their deadlines are considered to be successfully delivered. Packets received beyond these deadlines are considered lost.

To quantify the performance of a voice flow, we introduce a performance metric called the Packet Success Rate (PSR) that measures the percentage of voice packets that are successfully delivered to a voice application. That is, in addition to lost packets, the packet success rate excludes all packets that arrive but miss their deadline. The PSR value is the basic metric we use in this work for determining whether a single voice flow (or call leg) is supported.

Generally, voice calls contain both incoming and outgoing call legs. In a typical WLAN environment, which employs a centralized Access Point (AP), voice calls contain call legs in the uplink (terminal-to-AP) direction and the downlink (AP-to-terminal) direction. In such an environment, we consider a voice call to be supported if and only if both legs of the call have an acceptable PSR.

Based on this definition of a supported call, we define WLAN voice capacity as the total number of simultaneous calls that can be supported. More precisely, we define voice capacity to be the first point at which adding an additional call to the system results in at least one call not being supported with acceptable quality. In other words, the over-capacity is defined as the first point at which the system can no longer support the entire voice population. Defining capacity in this fashion ensures that all users in a WLAN operating within capacity limits are supported. Further, this definition naturally fits with notions of system fairness and can be easily integrated into typical admission control policies.

To quantify WLAN voice capacity we need to establish a delay budget for voice packets. There are two kinds of delays in telecommunication networks: medium propagation delays and handling delays. Medium propagation delays, being on the order of nanoseconds, are relatively insignificant when compared with handling delays and are not considered in our analysis. Handling delays are influenced by a variety of factors including, coding/decoding at codecs, endpoint packetization, core network traversal, and wireless network access. In our experimental setup we attempt to isolate the wireless network access delays which begin the instant a voice packet is handed to the MAC layer (at either the AP or a wireless station) and ends when that packet is received by the corresponding wireless MAC layer. This delay comprises both queuing delays at the MAC and medium access delays.

Based on the one-way end-to-end delay requirements of a voice flow, we calculate an upper bound for the tolerable wireless access delay for various networking environments. We use the worst-case scenario as our baseline, where both ends of the voice communication are connected to a wide-area network through wireless hops. Voice packets in such a conversation encounter the following delays: coding delays δ_c at the sending station, decoding delays, δ_c, at the receiving station, packetization delays, δ_p, at the sender, core network traversal delay, δ_{CN}, and wireless access delays, δ_{WN}, to access each wireless network. Thus, the one way end-to-end delay, Δ_{OW}, encountered by the packet may be expressed as shown in Eq. (1).

$$\Delta_{OW} = 2\delta_c + \delta_p + \delta_{CN} + 2\delta_{WN} \qquad (1)$$

We define $\tilde{D}$ as the maximum tolerable one-way delay or deadline by which a voice packet must arrive at its destination to be considered acceptable by the application. ETSI-TR 101 defines $\tilde{D}$ to be 200 ms. Then, an upper-bound for δ_{WN} can be calculated by replacing Δ_{OW} with $\tilde{D}$ and rearranging the terms in Eq. (1).

$$\delta_{WN} \leq \frac{\tilde{D} - 2\delta_c - \delta_p - \delta_{CN}}{2} \qquad (2)$$

In this study, we consider both Local Area Network (LAN) and Wide Area Network (WAN) scenarios. For the LAN analysis we assume no appreciable core network delay ($\delta_{CN} \approx 0$) and for the WAN analysis we assume a non-zero core network delay of $\delta_{CN} = 50$ ms.

4. EXPERIMENTAL TESTBED AND RESULTS

4.1 Testbed Setup and Experimental Methodology

We have developed a testbed for the purpose of studying voice capacity in IEEE 802.11b networks. The experimental testbed shown in Figure 3 contains 15 laptops that connect to a Cisco AP-350 access point using

Orinoco Gold PCMCIA cards. The AP connects the wireless network to a wireline network through a router.

The wireline network contains 4 desktops that act as correspondent hosts for voice conversations with the wireless hosts. Each voice conversation is established between one wireless and one wired participant. Wired hosts are capable of supporting multiple voice instances in software. Machines on the wired networks are inter-connected using 100 Mbps Ethernet cables. An NTP server is setup to provide time synchronization between all the machines in the testbed. We developed methods to ensure that the maximum time drift for any machine from the NTP server is kept less than 500 microseconds.

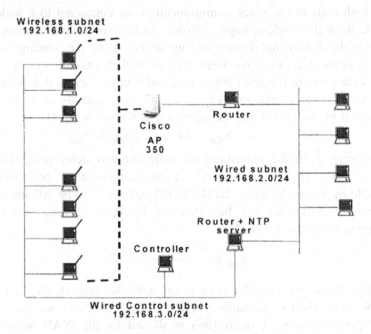

Figure 3 Experimental Testbed

All machines in the testbed have the RedHat 7.2 version of Linux running kernel version 2.4.7-10. The testbed machines contained a voice conversation model called VGEN. VGEN is an in-house voice generation tool that produces network traffic corresponding to conversational speech in compliance with ITU-T recommendation P.59 (ITU-T Rec. P.59). For further details on VGEN and the voice model, we refer the reader to (ITU-T Rec. P.59; Elaoud and Anjum 2003)

Our experimental goal is to isolate the MAC-layer and investigate the DCF's ability to resolve contention and packet collision to deliver quality real-time voice service. We are more concerned with the effects of medium contention than with signal degradation due to radio propagation. Hence, we operate our experiments in favorable radio environments where signal strengths are uniformly high and propagation delays are uniformly low for each terminal. This was accomplished by equidistantly placing the terminals within a 2-meter radius of the access point to ensure parity in terms of received signal strength and propagation delay.

All generated voice traffic involves a wireless and a wired host so that no traffic is generated between wireless hosts. VGEN parameters were chosen to emulate the G.711 codec with 10ms packetization intervals, 80 byte packets and silence suppression. Each voice "call" lasted for 3 minutes and consisted of bidirectional traffic. The number of voice users in the experiments was varied from 1 to 15. Each experiment run was repeated 5 times and results were average over all five runs.

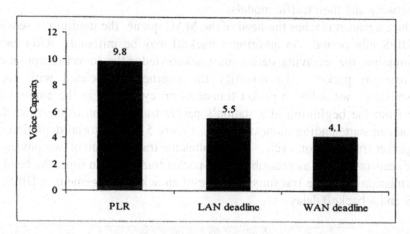

Figure 4 Measured Voice Capacity in IEEE 802.11b

4.2 Experimental Results

Figure 4 shows the number of supported voice calls in our experiments. The "PLR" label represents voice capacity determined solely based on packet losses, while the "LAN deadline" and "WAN deadline" labels represent voice capacities that also incorporate the appropriate deadline for the LAN and WAN scenario respectively. The figure shows that the network can only support about 10 voice conversations when delays are not incorporated. When packet deadlines are assumed, the capacity drops to less than 6 supported conversations in the LAN environment and about 4 supported calls in the WAN environment. This capacity is further reduced once data is added into the network. This small capacity represents a huge hurdle for WLAN providers who want to integrate voice into their public service offerings. An in-depth understanding of the limiting factors is of paramount importance to both equipment manufacturers and service providers. In the rest of this chapter we will provide a simple analysis to identify and quantify the voice capacity limiting factors.

5.   EFFECTS OF OVERHEAD ON CAPACITY

Each packet transmission attempt incurs a fixed packet header overhead and a variable access overhead due to the CSMA/CA mechanism. The header overhead depends on the traffic type and is fixed across all packets of the same type. However, the access overhead varies from packet to packet. The access overhead is contingent upon the number of contending stations in the network and their traffic models.

Once a packet reaches the head of the MAC queue, the medium is sensed for DIFS idle period. In addition a backoff may be initiated. After each transmission, the receiving station must acknowledge the correct reception of the received packet. To quantify the overhead associated with each transmission, we define a packet transmission cycle (T_c) as the amount of time from the beginning of a station's packet transmission to the time the station can start sending its next packet. Figure 5 is a pictorial depiction of the packet transmission cycle. T_c includes the transmission of two physical layer headers (known as preambles), the packet transmission time, the header transmission time, the transmission time of an acknowledgement, a DIFS, a SIFS and a backoff delay.

DIF	Backoff	PHY	Packet Transmission	SIFS	PHY	ACK

Figure 5 Packet Transmission Cycle

Let T_{phy} be the time to transmit the PHY header (or preamble), T_{ack} be the time to transmit a MAC layer ACK, T_b be the time taken by a backoff, T_{pl} be the time to transmit the packet payload, and T_h be the time to transmit the packet header overhead including all networking headers and MAC header. Then T_c can be expressed as follows

$$T_c = 2T_{phy} + T_{ack} + DIFS + SIFS + T_b + T_h + T_{pl} \qquad (3)$$

The first four terms in Eq. (3) represent a constant overhead for each packet transmission regardless of the packet type. The header overhead T_h, however, defers depending on the type of the transmitted packet. For instance for voice traffic T_h is the time to transmit 68 bytes of headers: 12-

byte RTP header + 8-byte UDP header + 20-byte IP header + 28-byte 802.11 header. Since we are interested in studying voice capacity in IEEE 802.11b networks, we will assume T_h to be a constant term. T_b is a variable backoff delay that hinges upon the number of contending stations and their traffic models.

Let T_f be the fixed overhead associated with each voice packet. T_f is independent of the payload size and the number of contending stations in the network. We define T_f as shown in Eq. (4).

$$T_f = 2T_{phy} + T_{ack} + SIFS + T_h \qquad (4)$$

We also define T_a as the variable access overhead that each packet experiences to access the channel. That is:

$$T_a = DIFS + T_b \qquad (5)$$

Then Tc can be expressed as follows:

$$T_c = T_a + T_f + T_{pl} \qquad (6)$$

From Eq. (6) we can calculate a user's transmission efficiency η as the ratio of the packet payload to the transmission cycle T_c as shown in Eq. (7)

$$\eta = \frac{T_{pl}}{Tc} = \frac{T_{pl}}{T_a + T_f + T_{pl}} \qquad (7)$$

In the next subsections we evaluate each component of the overhead and quantify its effects on the voice capacity in IEEE 802.11b networks.

5.1   Fixed Overhead

Generally, voice is sampled at fixed frequencies. For instance, G711 codec samples voice at either 64 or 56 kilo bits per second. After bits are constructed, they are grouped into packets. This is generally known as packetization. G711, for example, supports three types of packetization: 10-millisecond, 20-millsecond, and 30-millisecond packetization equivalent to a voice packet payload of 80 bytes, 160 bytes, and 240 bytes respectively. Each voice over IP (VoIP) packet incurs fixed packet header overhead. Specifically, each packet contains a 12-byte RTP header, an 8-byte UDP header, a 20-byte IP header, and 28-byte IEEE 802.11 header. In addition each MAC frame requires a physical layer header (preamble) of 192 bits and a 14-byte MAC acknowledgement. Regardless of the data transmission rate, preambles are transmitted at 1 Mbps and MAC acknowledgements are transmitted at either 1 or 2 Mbps.

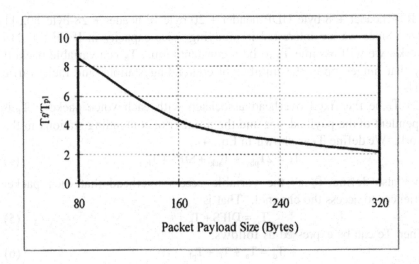

Figure 6 Fixed Packet Overhead

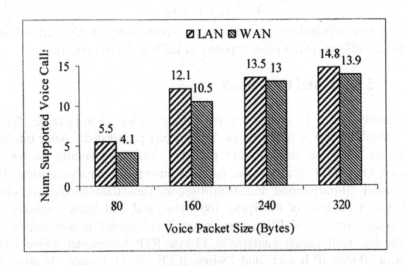

Figure 7 Effects of Packet Size on Voice Capacity

Figure 6 shows the ratio of the fixed overhead time to the transmission time of the packet payload. The figure shows that the overhead is twice to eight times the payload transmission time for practical voice packets of 80 to 240 bytes. This large overhead greatly contributes to the small voice capacity in IEEE 802.11b networks. Hence enlarging the voice packets and/or suppressing headers can improve voice capacity. Figure 7 shows experimental voice capacity measurements for various voice packet sizes in both the LAN and the WAN environments. We can see that the capacity more than doubles when the voice packet size is doubled from 80 bytes to 160 bytes. However, the increase in capacity tails off as the packet size is further increased. This suggests that the header overhead becomes less impacting as the payload size increases and that other overheads become the limiting factors for the voice capacity. In the next section we look at the access overhead and how it impacts the voice capacity in an IEEE 802.11b network.

5.2 Access Overhead in IEEE 802.11

As explained in Section 2, the backoff time depends on the contention window as well as on the number of users in the system. The backoff procedure of a station is suspended every time another station starts transmitting. Thus, if there is more than one station in the network, the backoff time may include several DIFS periods in addition to other stations' transmission duration. Thus the backoff time accounts for contention window count down and deferment during other users' transmissions. Therefore, Ta is a function of the number of contenders in the network, the contention window W, the slot size S, the time to transmit a packet including its header overhead $(T_{pl} + T_h)$, the time to transmit an 802.11 acknowledgement T_{ack}, DIFS, and SIFS. It is trivial that the minimum access delay is equal to DIFS which occurs if there are no transmitters during the first medium sensing

In IEEE 802.11, contention window is randomly selected from a uniform distribution in the range [0, W]. Hence the average contention window is ½W. The number of deferments that each station encounters depends on the chosen contention window, the number of contending stations in the network and their traffic models. The number of deferment can be modeled as random variable χ whose distribution depends on the number of contending stations and on their traffic models. Given the average contention window and the expected value of χ, $\bar{\chi}$, we compute the average access delay as shown in Eq. (8)

$$\bar{T}_a = \text{DIFS} + \frac{W}{2}S + \bar{\chi}(\text{DIFS} + T_p) \tag{8}$$

where T_p is the time to transmit a MAC frame and its acknowledgement as defined by Eq. (9)

$$T_p = 2T_{phy} + T_{ack} + \text{SIFS} + T_h + T_{pl} \tag{9}$$

The contention window W depends on the status of previous transmissions. In fact W is doubled after each collision and reset to its minimum (CW_{min}) after a successful transmission. In IEEE 802.11b the contention window W for any user is in the range of 31 to 1023. Figure 7 shows the ratio of the average access delay to the payload transmission time for W = 31 and for various user population. The figure assumes that number of deferments is a uniform distribution in the range [0, N], where N is the number of contenders in the network. The figure shows that for typical voice packets (80-240 bytes), the average access delay is more than 20 times larger than the payload transmission time. This implies that a station sending voice spends most of its time backing off rather than transmitting. The access delay constitutes the largest overhead and represents the bottleneck that limits the voice capacity in such a network.

6.   EFFECTS OF FAIRNESS ON CAPACITY

Generally, voice conversations contain uplink and downlink components. Each voice station contends for the wireless medium to transmit the uplink portion of its voice conversation. On the other hand, the access point must contend on behalf of all of the downlink voice components. That is, the access point contends for the wireless channel to transmit roughly half of the traffic. The access mechanism of the IEEE 802.11 is designed to provide equal medium access to all wireless stations, including the access point (ISO/IEC 1999). This access mechanism assures that if there are N contending stations in the network, in addition to the access point, each station, including the access point, obtains access to the channel with probability 1/(N+1).

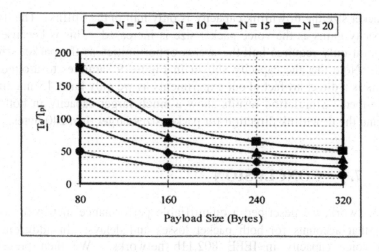

Figure 8 Channel Efficiency with Deferment

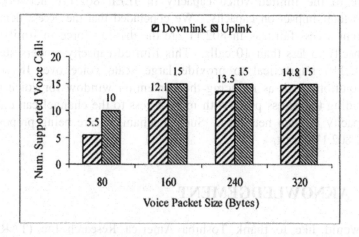

Figure 9 Effects of Fairness on Voice Capacity

For m two-way conversations, the access point contends on behalf of all m downlink voice flows. In order to obtain its fair share of the channel, the AP must be granted at least one-half of the wireless medium. However, the standard 802.11b DCF mode only grants the AP with about 1/m of the medium access opportunities. This access mechanism renders the access point a bottleneck resulting in poor voice performance for the downlink flows and thus limits the total voice capacity. Figure 9 shows the number of voice calls supported in the uplink and downlink directions. The figure shows that the downlink is the bottleneck of the network. At the smallest voice packet size the downlink capacity is only half of the uplink. The down link capacity gains as the voice packet size is increased. This is because the access point only requires half the access opportunities as the packet size is doubled. Note that the capacity shown in Figure 9 saturates to around 15 calls; this is a direct artifact of our experiments as we only have 15 machines in our testbed. Simulation results show a much larger capacity in both the uplink and the downlink direction for packet sizes larger than 240 bytes.

7.   CONCLUSION

In this work, we described a quantifiable performance metric for voice sessions that accounts for both packet losses and delays. In addition, we defined voice capacity in IEEE 802.11b networks. We then presented experimental voice capacity measurements. Results show that voice capacity is limited to about 10 conversations. We identified the factors that contribute to the limited voice capacity in IEEE 802.11b networks and quantified their impact on capacity. We concluded that the access delay and the inherent access fairness in 802.11 are the driving force in limiting the voice capacity to less than 10 calls. This limited capacity renders standard IEEE 802.11b impractical to provide large scale voice over IP service. Simple solutions such as reducing the contention window for voice traffic and providing the access point with more access to the channel can enhance voice capacity in such networks. Such mechanisms are being proposed in the IEEE 802.11e draft.

AKNOWLEDGEMENT

We would like to thank Toshiba America Research Inc (TARI) for supporting this work. We also would like to extend our appreciation to our colleagues David Famolari and Abhrajit Ghosh for their help to setup the testbed and collect some of the results used in this chapter.

REFERENCES

Akyildiz I. et al, "Medium access control protocols for multimedia traffic in wireless networks", IEEE Net. Mag., vol. 13, no 4, pp39-47, Jul/Aug 1999.
Chen D. Y., Garg S., Kappes M. and Trivedi K.S., "Supporting VBR VoIP traffic in IEEE 802.11 WLAN in PCF mode", in OPNETWORK'02, August 2002

Elaoud M, Anjum F., et al. "Voice Performance in WLAN Networks— An Experimental Study". Globecom Dec 2003.

Evers J., "City of Cerritos aims to be biggest U.S. Wi-Fi hot spot", IDG News Service, December 3, 2003

ETSI TR 101 329-6 V2.1.1, "Telecommunications and Internet Protocol Harmonization Over Networks (TIPHON)", Release 3; Part 1

Garg S., Kappes M., "Can I add a VoIP Call?" IEEE International Conference on Communications (ICC), 2003

IEEE STD 802.1lb-1999 (Supplement to ANSI/IEEE Std 802.11, 1999 Edition).

ISO/IEC and IEEE draft international standards, "Part 11: Wireless LAN Medium Access Control (MAC) and Physical Layer (PHY) specifications", ISO/IEC 8802-11, IEEE P802.11/D10, Jan 1999.

ITU-T Rec. P.862 "Perceptual Evaluation of Speech Quality (PESQ), an objective method for end-to-end speech quality assessment of narrow-band telephone networks and speech codecs." February 2001.

ITU-T Rec. G.107 "The E-Model, a computational model for use in transmission planning"

ITU-T Rec. P.59 Artificial Conversational Speech

Koepsel A. and Wolisz A., "Voice transmission in an IEEE 802.11 WLAN Based Access Network", In Proc. of WoWMoM 2001, pp. 24-33, Rom, Italy, July 2001

Kubbar and Mouftah H., "Multiple access voice protocols for wireless ATM: problems definition and design objectives", IEEE Communications Magazine, vol. 25, no. 11, pp93-99, Nov 1997

Nortel Networks, "Packet Loss and Packet Loss Concealment: A summary of how lost or late packets affect speech quality and how concealment is achieved." Technical Brief

SpectraLink Inc "Netlink Wireless telephones FAQ", http://www.spectralink.com/products/pdfs/Netlink\%20FAQ.pdf. Last visited on January 23, 2004",

Stine J. and deVeciana G., "Tactical communications using the IEEE 802.11 MAC protocol", Milcom pp 575-82, Oct 1998

Veeraraghavan M., Cocker N. and Moors T., "Support of voice services in IEEE 802.11 wireless LANs", IEEE Infocom 2000.

Chapter 9

ADHOC WIRLESS NETWORKING USING MOBILE BACKBONES

IZHAK RUBIN, ARASH BEHZAD, HUEI-JIUN JU, RUNHE ZHANG, XIAOLONG HUANG, YICHEN LIU, RIMA KHALAF
Electrical Engineering Department, University of California (UCLA), Los Angeles, CA 90095-1594, USA

Abstract: We introduce an adhoc wireless mobile network that employs a hierarchical networking architecture. The network nodes have different capabilities, and are thus divided into high and low capacity classes. We present a topological synthesis algorithm that selects a subset of the high capacity nodes to form a backbone network (Bnet). The latter consists of interconnected backbone nodes that intercommunicate across higher power (or regular) links, and may also make use of unmanned vehicles (UVs), including airborne UAVs orbiting at multiple altitudes, as well as ground based UGVs, to form a multi-tier backbone. Each backbone node controls the allocation of communications resources associated with client nodes that reside in its managed cluster of nodes (forming its Access Net - Anet). We introduce the Mobile Backbone Network Protocol (MBNP) to implement the key networking schemes for such a Mobile Backbone Network (MBN). Our description of this protocol involves the following procedures: backbone network topological synthesis; on-demand and/or proactive routing mechanisms; power control based MAC layer protocols; and network/MAC cross-layer resource allocation schemes. The MBNP serves to allocate resources across the network to ensure that user applications are granted acceptable quality-of-service (QoS) performance, while striving to ensure a highly survivable and robust backbone-oriented networking architecture. We include in MBNP a new class of on-demand routing algorithms, identified as MBNR, that employ the backbone network for selective forwarding of route-request messages, while striving to achieve an efficient MAC layer operation. We enhance these new routing algorithms by incorporating link stability estimates to attain a robust routing operation. For this purpose, links that are determined by a node to be in an unstable state are dynamically eliminated from the backbone subnetwork that is used for establishing new routes. To ensure service quality for admitted flows, including the attainment of low delay jitter performance levels for supported realtime streams, we introduce flow admission control mechanisms into our MBN based on-demand routing operation. To further enhance the operation of the network, and to achieve a more stable backbone system, when required, we incorporate the use of unmanned vehicles (UVs), including UGVs and UAVs. We present new spatial-reuse based power control algorithms for efficient utilization of the net MAC resources through the use of time slot allocations, as well as through the use of CSMA/CA based (IEEE 802.11 type) protocols.

1.   INTRODUCTION

We have recently been investigating the operation of mobile wireless networks through the embedded establishment of a Mobile Backbone Network (MBN). In ([1]-[5], [13], [16]), we have presented objectives and methods for the design of MBN topological layouts. A mobile backbone network consists of a backbone network (Bnet; or a multiple number of Bnet components), access nets (Anets), and backbone-less oriented (flat) adhoc network(s). Its structure is illustrated in Figure 1. Large solid circles (nodes) and the thick solid lines connecting them to each other represent the Bnet. Dashed ovals consisting of thin solid lines connecting small solid circles (nodes) represent the Anets. The small solid circles (nodes) and the thin dashed lines connecting them to each other represent the flat adhoc subnetwork. In effect, by requiring each non-backbone node to select a particular node on the backbone network, identified as its associated backbone node, and by allowing a non-backbone node to reach its associated backbone node in multiple hops, provided such a path is available, non-backbone nodes can generally be affiliated, with an Anet and its managing backbone node. The MBN is designed so that it involves a sufficient but not excessive number of backbone nodes, while providing high coverage, so that high fraction of the lower capacity nodes (excluding outlier groups) can access at least a single Backbone Node (BN) through a path of at most h hops. The MBN strives to implement a survivable/robust Bnet topology, such as that realized by a k-connected backbone network.

We consider an adhoc wireless network in which some nodes are outfitted with higher capability resources and mechanisms that make them more attractive as backbone capable nodes (BCNs). The role of a backbone node can be served by a wide range of specially designated nodes that have advantageous capabilities involving attributes of: communications-access and geographical coverage (as is the case for orbiting, roving or guided airborne or ground based unmanned vehicles); platforms with designated organizational or functional roles (headquarters or command posts), and other stationary relay nodes or mobile vehicles characterized by higher communications and processing capacities. In turn, a small sensor node is highly power limited and is not expected to perform demanding processing and communications operations; it is thus effective for its communications flows to be managed by a backbone node. A node that is not a candidate for selection to perform as a BN is classified by us as a Regular Node (RN). The Bnet is dynamically formed to consist of interconnected BNs. Nodes can communicate to each other either through a flat multi-hop adhoc networking route, or through the Bnet. Under the latter approach, the nodes form a hierarchical (and multi-tier based using UGVs, UAVs and satellite platforms) communications networking mobile infrastructure, which is particularly attractive when demanding the provision of a communications route that guarantees a flow distinct quality of service (QoS) performance.

The remaining of this paper is structured as follows. The topology synthesis protocol is described in Section 2. We present an MBN-based on-demand routing protocol in section 3. In section 4, we describe the operation and exhibit performance features of our combined MBN routing and flow control structure. The role of UGVs in aiding the MBN protocol is demonstrated in section 5. Integrated scheduling and power control schemes are discussed in section 6. A power controlled CSMA/CA medium access control scheme is presented in section 7.

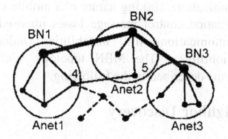

Figure 1 The decomposition of a mobile backbone network into a Bnet, three Anets, and a regular adhoc network with h = 1 and k = 1.

2. TOPOLOGY SYNTHESIS

The mobile backbone networking architecture as employed herein has been introduced by Professor Izhak Rubin of the UCLA Electrical Engineering department around 1995. In [1] – [5], we present the overall concept, operation and characteristics of the Mobile Backbone Network (MBN) that constructs a multi-tier hierarchical architecture for wireless mobile adhoc networks. In a Mobile Backbone Network (MBN), nodes are classified into *Backbone Capable Nodes (BCNs)* and *Regular Nodes (RNs)* based on their respective computation, processing, power and transmission capabilities. Under the MBN protocol, a *Backbone Network (Bnet)* is formed by dynamically electing Backbone Nodes (BNs) among BCNs. In this section, we present a distributed MBN backbone topology synthesis algorithm that assumes each node to have a single radio and all nodes to operate on the same frequency band. BCNs are superior to RNs in the sense of packet routing and forwarding ability and computational capability. We note that the methods described herein are readily applied when multiple frequency bands (or time slots, or CDMA codes) are employed to avoid interferences between Anets and Bnets, as well as between neighboring Anets and between different backbone tiers. For example, under our TBONE implementation described in [5], Anets coordinate their MAC operations in a time slotted fashion to avoid interferences, while Bnets may employ distinct frequency bands as well as use directional antennas.

The MBN topology synthesis algorithm is fully distributed. Every node makes decisions independently by using local information available to it. In this manner, network-wide global information is not required for constructing the backbone network layout. Furthermore, the MBNP version presented here is not based on assuming a perfect MAC layer operation. We design the topology synthesis mechanism to incorporate into its operation the unreliable communications linking nature of a mobile wireless network, and take into consideration control message losses imposed by MAC layer collisions and communications link instabilities (induced by fading, interferences and mobility). The MBN topology synthesis algorithm includes the components described in the following.

2.1   Neighbor Discovery

Every node periodically sends out a hello message that contains the node ID, node status (BN, BCN, RN), nodal weight, associated BN ID (if any), and BN neighbor list. Also, every node updates its neighbor list periodically based on the observation of hello messages received from each of its neighbors over a sliding window period. Through such a neighbor discovery process, each (BN or BCN type) node learns the network inter-nodal link connectivity in its 2-hop BN neighborhood as well as keeps record of its 1-hop BCN and RN neighbors.

2.2   Association Algorithm

Every unassociated node that is in the BCN or RN state attempts to associate with a BN. It will strive to identify, among its neighbors, the BN that has the highest *Weight* to associate with. The *Weight* of a node can be based on its ID, degree (i.e., number of its BN node neighbors and possibly also number of current associated client nodes), processing/communications capability, congestion level, and a nodal/link stability measure (indicating the robustness of its attached communications links and its own degree of mobility and reliability). A weight vector is employed when different weights are stated for different traffic classes and flow/message priority levels, including QoS preservation admission controls. Using the weight function, we offer the network designer the flexibility to impose various features and constraints in dynamically constructing the backbone layout. If no acceptable neighboring BN is detected, the node attempts to find a BCN, selecting among all its neighboring BCNs, including itself, the one with the highest weight. If such BCNs are not available, the node attempts to find a RN with the highest weight, among its RN neighbors. The latter RN is used as a relay node for reaching (when feasible) a BN with whom the RN elects to associate, becoming a member of this BN's Anet. The latter neighboring RN will act as its predecessor node on the multi-hop route leading from the managing BN to the underlying RN.

2.3 BCN to BN Conversion Algorithm

Backbone Capable Nodes (BCNs) can be either in the BN state (and then identified as BNs) or BCN state. There are two main criteria for a BCN to convert itself to a BN. The first one is for maintaining the connectivity of the backbone network; i.e. a BCN node finds that by converting itself to a BN it will upgrade the Bnet connectivity. The second one is for client coverage: a BCN that receives an association request from a BCN or RN, converts itself to a BN.

2.4 BN to BCN Conversion Algorithm

Under wide-scope and extensively high nodal mobility in the region of operations, it is possible for many nodes to eventually become BNs. The BN to BCN conversion mechanism serves to reduce the number of redundant BNs, while maintaining network connectivity and coverage. For a BN to convert to a BCN, we invoke the following conditions on the BN: (1) All of its BN neighbors have at least one common BN neighbor whose weight is higher than the weight of the underlying BN that is considering to convert. (2) Each of its BCN clients have at least one other BN neighbor.

3. MBN-BASED ON DEMAND ROUTING

The mobile backbone layout provides a reliable infrastructure for MBN based routing protocols. On-demand routing protocols for adhoc networks such as AODV and DSR, instruct a source that initiates a flow to discover a source-destination route across the current network topology. For this purpose, the source node broadcasts (floods) route request (RREQ) packets across the entire network. When the request packet (or one of its replicas) reaches its destination, a reply is returned and the route is set. Clearly, this by itself is not a scalable approach. As the network size grows, or the number of active sources increases, or when the network contains a high density of nodes, the high traffic intensity caused by the control overhead, induced by the high rate of RREQ broadcast messaging flows, can result in excessive demand imposed on link capacity resources, leaving insufficient residual capacity for data packet support. In addition, high control packer rate also induces severe MAC contentions for access to the shared multiple-access radio channels, leading to a high rate of retransmissions and packet discards.

3.1 MBNP On-Demand Routing (MBNR)

Under the MBN protocol presented above, a backbone network (Bnet) is dynamically formed. Assume for presentation purposes that the formed Bnet is a connected network. Assuming that the graph that consists of all network nodes (whereby two nodes are set to be connected if they can communicate successfully to each other) is connected, each RN and BCN is associated with an elected BN, having a single-hop or multi-hop path connection to this BN. Each BN keeps a registry of its clients. The registry is updated as registration messages are received. The MBN oriented routing algorithm to be employed can be proactive or reactive. In this paper, we focus on the interaction between on-demand routing protocols and the MBNP, presenting an MBNP based on-demand routing protocol (identified as MBNR). The latter is shown to provide a scalable and robust solution to mobile adhoc networking.

We use AODV as an example to illustrate the operation and performance improvement attained by the MBNR protocol. The MBNR version presented here extends and modifies an AODV type operation, and is thus also identified as MBNP-AODV, and it works as follows. Upon discovering a route to a destination node, a BCN/RN source node floods a route request (RREQ) along up to $h+k$ hops to reach the Mobile Backbone. The hop count h is obtained from the topological synthesis algorithm (Section 2) and it indicates the hop distance between the source node and its associated BN.

(BN control/beacon messages are broadcasted in its Anet, so that each client node can confirm its continued association.) Note that a margin (*k*) of hops has been added to allow access to nearby BNs in case the associated BN has just moved and no new associations have been yet been formed. When a route request reaches the *Mobile Backbone,* i.e. the source node's associated BN, or any other BN, the route request is flooded within the *Mobile Backbone* until a route request is received by the associated BN of the destination node. Upon receipt of this route request, the latter destination BN floods the route request within a distance of (h'+k) hops (covering its Anet, as well as a nearby zone, in case the node has somewhat moved and did not yet have the chance to re-associate) to reach the destination node. Note that h' indicates the hop count between the destination node and its associated BN. We note that to cover multicast routing, or to deliver a message to a group of nodes that are associated with multiple BNs, the invocation of the RREQ based route discovery process actually results in the discovery of a delivery Bnet subnetwork (such as a multicast tree).

This discovery process ensures the source node that it will find a route to the destination since the MBNP provides a connected backbone network that offers a access to each network node (recalling that the overall network is connected). This route may be somewhat longer than the one that the source may obtain if it were permitted to flood its RREQ packets across the whole network. However, the selective forwarding process, in using only the backbone to discover routes, reduces routing overhead and leads to a scalable operation. Furthermore, it tends also to substantially reduce MAC contentions. Under this protocol, if the destination node is at a distance that is equal or less than *h* hops away from the source node, it replies directly to the source node, as performed under AODV, without resorting to the selective forwarding process enacted by the backbone nodes. In turn, messages produced by a source node to a destination that is located *h+k* hops or more away, are transported through the backbone network.

For illustrative purposes, we consider in this section a special network scenario under which all the nodes in the network are Backbone Capable Nodes (BCNs). Note that in this case, all the client nodes of a BN are BCNs that are 1-hop away from it. No client registry is required to be kept at a BN for routing purposes, since it recognizes all nodes that it can reach by examining the continuously updated list of its nodal neighbors. The operation of MBNP-AODV in this special case is thus based on the key rule that permits only BNs to forward route request packets (RREQs).

To demonstrate the performance of the MBN based topology synthesis algorithm and routing protocol presented above, we carry out the simulation analyses described in the following. We randomly place 100 nodes in 1000 m x 1000 m and 1250 m x 1250 m areas. We assume an IEEE802.11b MAC operation at a data rate of 2 Mbps. We compare the throughput performance of AODV and MBNR (also identified as MBNP-AODV) protocols. The moving speed ranges from 0 to 12 m/s. Packet inter-arrival times are exponentially distributed with an average of 0.5 sec. The packet size is set to 256 Bytes and 50 end-to-end traffic flows load the network. The total traffic rate offered to the entire network is thus equal to 205 Kbps. This represents a relatively light network offered loading level, which enables us to focus on the impact of nodal mobility on network performance.

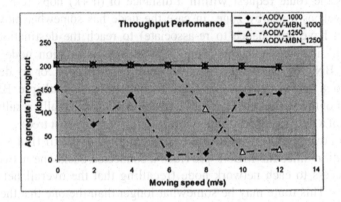

Figure 2 Throughput performance of AODV and MBNR (MBNP-AODV).

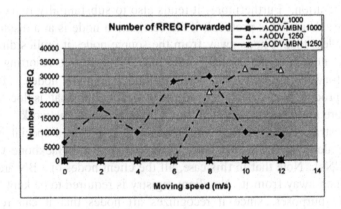

Figure 3 Average number of RREQ forwarded by each node.

Nodal mobility can cause an existing route to break. In this case (assuming no local repair is attempted or is successful), the source node needs to initiate another RREQ packet to discover a new route. As the nodal mobility level increases, additional RREQ packets will be generated and flooded across the network. Such an overload of control traffic will ultimately cause the network data throughput to deteriorate sharply, being further aggravated by intensified MAC contention based collisions induced by the increased control messaging traffic loading the network, as shown in Figures 2 and 3.

3.2   MBNP-Robust QoS Routing

To enhance the robustness of the routes allocated to flows in the adhoc wireless network, we have studied the inclusion of link stability measures. The instability of a link is defined in terms of the probability of link breakage. Under our MBN protocol, each node uses a sliding window period to record a count (accumulated across the window interval) of the number of hello messages received from its neighbors. A larger count value is indicative of a more stable link. These link stability indices are used by each node in determining its weight index indicative of its suitability to serve as a backbone node. For routing purposes, the link stability measures are employed by a node in determining whether to include a link in its route discovery process. In this manner, unstable links are avoided. When a flow is allocated a route, we attempt to assign to it a path that will satisfy its QoS requirement, ensuring statistically an end-to-end path robustness requirement. Such an operation reduces the rate at which new routes must be discovered for a given flow, contributing to control overhead reduction for on-demand routing algorithms, and thus yielding enhanced throughput performance. This aspect of our routing protocol is identified as its robust QoS routing behavior.

Such robust QoS routing element can be incorporated into the operation of various on-demand routing algorithms, including AODV, DSR, and MBNR. When the source initiates a flow, it specifies its desired path stability requirement as a parameter included in its route request. Each intermediate node that is permitted to forward route request packets, upon reception of a route request packet, first calculates a cumulative end-to-end stability measure for each one of its outgoing links. The latter measure represents a stability index of the path traversed hence-to-forth by the request message, including the selected outgoing link. The node uses the updated cumulative measure to replace the request's current cumulative stability measure. When then attempting to decide on forwarding a request packet across a selected link, a forwarding node examines the packet's cumulative stability measurement. It forwards the route request to only those neighbors that are located across the links that keep the cumulative stability measure value higher than its specified lowest allowable level. In this manner, the request flooding mechanism automatically avoids unstable links, leading to reduced control overhead. As noted above, the integration of the robust QoS routing element described above into MBNP serves to select a more robust and stable backbone network, and thus to reduce the overhead involved in backbone network reconfigurations. It also ensures the rapid pruning of unstable communications links. We use it to realize a QoS based operation, in providing particularly stable routes to those flows that are more sensitive to route failures.

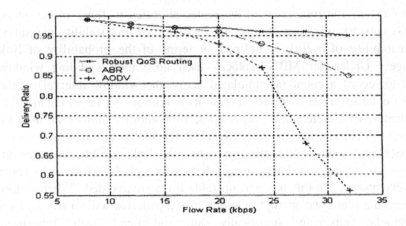

Figure 4 Illustration of the delivery ratio vs. flow rate.

We compare our robust QoS routing protocol, as described above, with two other protocols: 1. AODV and 2. Associative Based Routing (ABR) [12]. The latter operates in a manner similar to AODV but uses a link robustness index (in addition to the path's hop length) as a parameter to aid the decision made at the destination node in selecting the best route. Our simulation results, as illustrated in Figure 4 by the shown packet delivery ratio vs. loading rate performance curves, well demonstrate the capability of the robust QoS routing algorithm to yield significantly enhanced network performance. By eliminating the use of unstable links, the robust routing element achieves a much reduced routing overhead, increases the package delivery ratio, leads to the attainment of much higher throughput level, offering a more reliable and survivable network operation.

4.   MBN ROUTING WITH FLOW CONTROL

To ensure the performance of admitted flows across the network, we incorporate into the MBN routing system a flow control mechanism, identified as Mobile Backbone Network Routing with Flow Control (MBNR-FC). The MBNR-FC makes use of the embedded signaling mechanism that is established by an on-demand routing structure and that is used as part of the route discovery process. This mechanism is employed for the implementation of QoS based flow admission control procedure that is used to protect congested network zones by admitting and navigating packet flows across less congested areas.

Under MBNR-FC, each BN monitors its own congestion status by recording it's backlogged (network layer) packet queue size. (To regulate access by traffic/service classes and by message/flow priority levels, we keep record of the queue size levels of separate queues used to store packets by their corresponding categories.) If the queue size level surpasses a prescribed threshold, the BN stops relaying route requests packets. Each BN includes its congestion status in its periodically sent "hello" messages, thus informing its neighbors about its congestion state. Our study shows this mechanism to be highly effective for regulating admission of flows into the backbone network at congested access BNs. Furthermore, this flow control procedure is also effective in preventing neighbors of a congested BN from relaying request packets, thus acting to further reduce congestion induced performance deterioration.

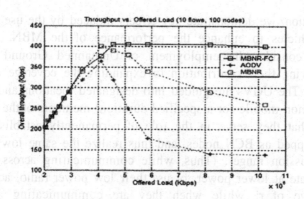

Figure 5 Throughput-Offered Load curve comparison.

Under our MBNR-FC scheme, we differentiate admitted flows from newly generated flows. A newly generated flow starts a new route discovery process, disregarding previously obtained routes. A flow attempt to discover and establish a route is terminated after a specified maximum number of unsuccessful attempts. In this manner, the flow control process protects the performance provided to admitted flows. In Fig. 5, we show that the MBNR-FC method significantly improves the throughput versus offered load performance behavior, when compared with AODV routing protocol and with MBN routing protocol that does not employ the underlying flow control scheme. In Fig. 6, we show that the MBNR-FC algorithm yields a significant reduction in the packet delay jitter performance of admitted flows, under relatively high traffic loading conditions.

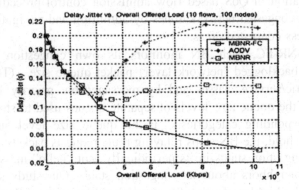

Figure 6 The delay jitter - offer load curve comparison.

5.   USE OF UNMANNED VEHICLES

In this section, we demonstrate key effects induced by the use of guided unmanned vehicles to enhance the performance of the MBN. For this purpose, we consider the employment of Unmanned Ground Vehicles (UGVs), showing their contribution to expanding the coverage of regular nodes (RNs). The UGVs are guided into designated locations and are used as stationary (non-mobile for a specified duration) members of the backbone network, so that they serve in the role of permanently established BNs. They are equipped as BCN nodes, and thus realize the same low and high power transmission ranges. (Thus, while communicating across the Anet they can operate at lower power, or use their low power radio, achieving a coverage range of r; while when they are communicating across the backbone, they may operate at higher power level, attaining a communications range of value R > r.) As seen in Fig.7, the introduction of the four UGVs (symmetrically located here within the area of operations of 1000 m x 1000 m), significantly improves RN coverage (as expressed by the displayed performance measure that represents the fractional number of covered (associated) RNs), especially when the low power transmission range *r* is larger. A key factor contributing to this enhancement is the large aggregate coverage area offered by the four UGVs realized by their non-overlapping individual coverage zones.

To guarantee a communications path for packets that are transported between stations located in distinct Anets, the interconnecting Bnet must be configured to form a connected network. This can be accomplished by either increasing the number of BCNs or increase the high power link transmission range. When the network contains an insufficient number of BCNs, the introduction of UGVs and their placement in advantageous locations, can lead to a substantial upgrade in the probability of a connected backbone network.

To demonstrate the latter feature, we carry out an MBN system simulation over the same area of operations as assumed above, further setting: number of regular nodes (n_r) = 200; *number of BCNs* (n_c) = 20; *number of UGVs* (v_{max}) = 4; *r* = 128 m. We consider two cases: an MBN that employs UGVs as well as one that does not contain them. In Fig. 8, we display the variation of the backbone network connectivity factor as a function of the high power transmission range parameter *R,* for both cases. As expected, the Bnet is more likely to be connected under longer communications ranges *R.* As we introduce UGVs into the network, we realize great improvement in connectivity. For example, for *R* = 192 m, when no UGVs are used, the connectivity factor (the probability that the backbone is connected) is equal to 58%, while when four UGVs are employed, the connectivity level is upgraded to 88%.

We note that even larger enhancements can be attained through the use of UAVs that orbit the area of operations at designated altitudes in a specified configuration. We have employed multi-tier constellations of UAVs (and satellites) to form an effective backbone that enhances the network connectivity, upgrades robustness to mobility and to link/nodal failures, as well as add capacity resources over overloaded spatial segments.

We have also studied the optimal location of UGVs and UAVs that can be guided into proper positions to form an upgraded connected backbone network. In [16], we have modeled such an optimization problem as a quadratic programming problem with quadratic constraints. We have introduced an efficient two-phase heuristic algorithm for the synthesis of the connected backbone. In the first phase, we form clusters of nodes (Anets) that are within communications range and place UVs as BNs within each cluster. In the second phase, we adjust the locations of the BNs, and add BNs, if required, to construct an efficient connected backbone network.

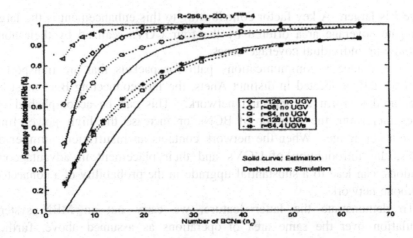

Figure 7 RN Coverage as a function of number of BCNs.

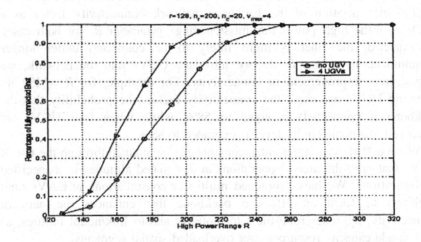

Figure 8 Backbone connectivity as a function of R.

6. JOINT SCHEDULING AND POWER CONTROL

We have developed and investigated a wide range of algorithms for integrated scheduling and power control using a central scheduling manager ([6], [7], [10]). Such schemes can be directly employed in the MBN Anets, noting that each Anet is managed by its BN. In the process of designing these algorithms, we have utilized our probabilistic analysis on the performance of graph-based scheduling algorithms [9] and our theoretical results on the impact of power control on the throughput capacity of adhoc wireless networks [11]. In the following, we succinctly describe one of our joint scheduling and power control algorithms (i.e., PCSA), which can specifically serve as a MAC mechanism for Anets. Details can be found in [7]. An extension of the algorithm to general (including Bnet oriented) topological configurations is described in [8].

Under the Power Controlled Scheduling Algorithm (PCSA), the backbone node (BN) instructs the Anet nodes to make power control adjustments while simultaneously allocating to them time slots to be used for transmission of their packets, per their requests (or dynamically estimated activity requirements). Our mathematical and simulation-based results indicate that our efficient graph coloring based approach can solve the joint power control and scheduling problem in a computationally effective manner. This algorithm, in contrast to other employed conventional graph based scheduling algorithms, satisfies the requirement that a minimum signal-to-interference and noise ratio (SINR) is met at all intended receivers. Our purpose in devising this algorithm is to achieve a higher net throughput level by attaining a high ratio of the average spatial reuse factor and the average path length. We show the PCSA scheme to lead to significant increase in the network throughput level through spatial reuse of the communications resources while (strongly Pareto) optimizing power consumption. While the PCSA scheme employs demand-assigned *STDMA (spatial TDMA) link scheduling,* the same analysis is readily adapted to *STDMA broadcast scheduling, spatial frequency division multiple access (SFDMA) broadcast scheduling,* and *SFDMA link scheduling,* as well as for the use of a SCDMA based approach.

PCSA consists of three major steps: synthesis of the Power-based Interference Graph, finding a maximal independent set of this Graph, and the power control operation.

6.1 Synthesis of the Power-based Interference Graph

PCSA's operation is based on the notion of the *Power-based Interference Graph* whose vertices are the transmission requests and an edge between two vertices indicates that the associated transmissions cannot be simultaneously received successfully under any (feasible) power allocation. We have derived, based on specified scenario information, sufficient and necessary conditions for the existence of an edge between two vertices. We show that the independence and chromatic numbers of the Power-based Interference Graph provide fundamental bounds on the optimal performance attained by a solution to the integrated scheduling and power control problem.

6.2 Finding a Maximal Independent Set

PCSA utilizes the *Minimal Degree Greedy Algorithm* (MDGA) to find a maximal independent set (transmission scenario) of the Power-based Interference Graph. We note that based on the properties of the Power-based Interference Graph, every subset of a maximal independent set with cardinality equal to two is a feasible transmission scenario. However, considering the accumulative effects of interference, the entire maximal independent set is not necessarily a feasible transmission scenario.

For the sake of simplicity, we have limited our approach for finding the maximal independent set to MDGA. As other variations, we may apply other techniques for finding a maximal independent set of the Power-based Interference Graph, such as column generation, genetic algorithm, semi-definite programming, or the algorithms introduced in [6]. We note that MDGA is a simple algorithm, which has been proven to be much better than previously claimed. In particular, it has been proven that MDGA achieves a performance ratio of $(\Delta + 2)/3$ for approximating independent sets in graphs with degree bounded by Δ. Moreover, MDGA almost always yields a solution characterized by a performance value that is at least half the independence number of a random graph.

6.3 Power Control Operation

In this step, PCSA strives to find the largest feasible transmission scenario that is a subset of the underlying maximal independent set. This problem is known to belong to the class of NP-hard problems. Therefore, based on the Perron-Frobenious theorem, we design the following greedy approach to find a near-optimal solution for the problem.

We check the feasibility of the derived maximal independent set by solving a system of linear equations. If the solution (i.e. the power levels) is feasible (which can be also verified a priori by examining the spectral radius of the associated matrix), the involved transmission requests are assigned to the underlying slot and are removed from the residual Power-based Interference Graph. Otherwise, one transmission request (for example the one that causes the maximum interference) is removed from the independent set and the transmission power vector is recalculated for the new system of linear equations. This procedure repeats until the remaining transmission requests form a feasible transmission scenario. This process converges after a maximum of M iterations, where M is the cardinality of the (initial) maximal independent set. Our simulation results indicate that in many cases the initial independent set, with no or very few transmission removals, forms a feasible transmission scenario.

The computational complexity of PCSA is dominated by finding a feasible transmission scenario at the latter step. Since solving a MxM system of linear equations has a complexity of $O(M^3)$, and the size of the maximal independent set is no more than $n/2$, it can be shown that the complexity of PCSA is less than $O(m^4 T)$, where m and T represent the number of nodes in the Anet and the number of the time slots in a timeframe, respectively. This is a reasonable level of complexity for an algorithm that solves such a demanding problem.

In Fig. 9, we compare the throughput performance attained under the use of the PCSA algorithm with that achieved by the graph-based scheduling algorithm (GBSA) [9] for a random Anet topology with 30 nodes. Under GBSA, no power control operation is conducted, and every node is set to transmit at a fixed power level P_{max} to guarantee its association with the BN. Consequently, assuming an interference range that is twice the communication range, no parallel transmissions are allowed in the Anet under the GBSA algorithm (noting the diameter of the Anet to be equal to at most two). However, based on the PCSA algorithm, a significant upgrade in the throughput level is exhibited in Fig. 9.

We note that PCSA is also a suitable scheme for Bnets, in particular for the purpose of supporting QoS, particularly when the total number of BNs in the Bnet is moderate.

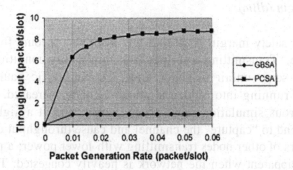

Figure 9 Illustration of the throughput upgrade attained under the PCSA algorithm for an Anet with 30 nodes.

7.   MEDIUM ACCESS CONTROL: CSMA/CA

We aim to improve the throughput-delay characteristics of the Anet by controlling the transmit power. It has been pointed out that an effective operational mode in wireless multihop networks is to transmit at the lowest power at which the network still remains connected. This strategy, however, does not seem to enhance throughput-delay characteristics when compared with approaches that employ higher power transmissions, when considering typical access or regional networks that may support a few hundred nodes. The main culprit in this matter stems from the dominance of the (increased) average hop count over the spatial re-use factor attained when one transmits at lower power levels. Furthermore, longer path multi-hopping introduces additional packet queuing delays at intermediate nodes as it increases the amount of internal traffic.

In contrast, we present the following approach. In an attempt to decrease the average hop count to reach the destination, we opt to adjust the transmit power level in accordance with the *per-link-minimality condition,* i.e. employing just enough power to reach the intended destination over the hop, or alternatively in a minimal number of hops, as dictated by terrain topography (and the maximum available transmit power level).   For an access net (Anet), or even a regional net with spanning ranges of hundreds of meters, such shorter paths are readily implemented.   Thus, a node i wishing to initiate transmission to a node j (in the same Anet) should transmit at the power level:

$P_{minimality}$ *(i) (dBm) = Reception-Threshold (dBm)+ Path-loss (i, j) (dBm) + Safety-Margin (dBm).*

The power safety margin is set to a few dBm's to account for the effects of interference. Transmitting at the per-link-minimality condition, however, gives rise to some unfair behavior under certain traffic patterns, as the probability of running into hidden terminals may be increased. As noticed by our numerous simulation runs, nodes transmitting at a higher transmit power level tend to "capture" the channel and cause throughput deterioration at the receivers of other nodes transmitting with lower power; a problem that is especially apparent when the network is heavily congested. This is a key problem that our fair distributed power control algorithm, described below, strives to tackle.

Our algorithm provides a power control modification to the CSMA/CA MAC protocol implemented in accordance with the IEEE 802.11 Standard. To implement our algorithm, we modify the header of the data packet to include the transmit power level.   Thus, every node overhearing a data packet can approximate its incurred path-loss by subtracting its received power level from the value of the transmit power specified in the packet's header. Also, every node in the network monitors the medium's activity over a recent sliding time window and records the fraction of time that the medium is busy. If the congestion is above a specified threshold level, and node i that is transmitting at a low power level hears a node k that is transmitting at higher power, node i increases it's transmit power to a level that is sufficient to activate carrier sensing at node k. The transmission power needed for carrier-sense-activation at another transceiver is calculated as:

$$P_{csa}(i) = \text{Carrier-Sense-Threshold} + \max_k \text{ path-loss}(i, k) + \text{safety-margin}$$

Thus, under high congestion conditions, a node transmits at max $(P_{minimality}, P_{csa})$. As demonstrated by the delay-throughput performance curves shown in Fig. 10, our distributed power control algorithm (DPC) achieves a significantly enhanced performance, when compared to schemes that use low power multihop transmissions (LOW) and to IEEE802.11 (Regular) based operation that employs no power control. Our scheme also exhibits significantly enhanced fairness behavior.

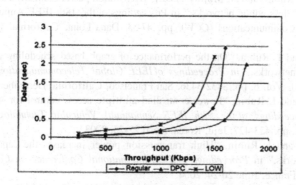

Figure 10 Throughput-delay characteristics, 50 nodes, 20 flows.

ACKNOWLEDGEMENTS

This work was supported by Office of Naval Research (ONR) under Contract No. N00014-01-C-0016, as part of the AINS (Autonomous Intelligent Networked Systems) project, and by the National Science Foundation (NSF) under Grant No. ANI-0087148.

REFERENCES

[1] I. Rubin and P. Vincent, "Topological synthesis of mobile wireless networks," in *Proceedings of IEEE MILCOM Conference,* session on Digital Battlefield Communications, Los Angeles, CA, October 2000.

[2] I. Rubin, X. Huang, Y. Liu and H. Ju, "A distributed stable backbone maintenance protocol for adhoc wireless networks," in *Proceedings of IEEE Vehicular Technology Conference (VTC Spring 2003),* Seju, Korea, April 2003.

[3] I. Rubin and P. Vincent, "topological synthesis of mobile wireless networks for managing adhoc wireless networks," in *Proceedings of IFIP/IEEE International Conference on Management of Multimedia Networks and Services (MMNS 2001),* DePaul University, Chicago, IL, October 29 - November 1, 2001.

[4] I. Rubin, R. Zhang and H.Ju, "Topological performance mobile backbone based wireless adhoc networks with unmanned vehicles," in *Proceedings of Wireless Communications and Networking Conference (WCNC'03),* New Orleans, LA, March 2003.

[5] I. Rubin, A. Behzad, R. Zhang, H. Luo and E. Caballero, "TBONE: A mobile-backbone protocol for adhoc wireless networks with unmanned vehicles," in *Proceedings of IEEE Aerospace Conference,* Vol. 6, pp. 2727-2740, Big Sky, Montana, March 9-16, 2002.

[6] I. Rubin and A. Behzad, "Cross-layer routing and multiple-access protocol for power-controlled wireless access nets," in *Proceedings of IEEE CAS Workshop on Wireless Communications and Networking,* Pasadena, California, September 5-6, 2002.

[7] A. Behzad and I. Rubin, "Multiple access protocol for power controlled wireless access nets," *IEEE Transactions on Mobile Computing,* Vol. 3, 2004.

[8] A. Behzad, I. Rubin, and A. Mojibi-Yazdi, "Distributed power controlled medium access control for wireless adhoc networks," in Proceedings of the 18th IEEE Annual Workshop on Computer Communications *(CCW),* pp. 47-53, Dana Point, California, October 20-21, 2003.

[9] A. Behzad and I. Rubin, "On the performance of graph based scheduling algorithms for packer radio networks," *in Proceedings of IEEE Global Telecommunications Conference (GLOBECOM),* Vol. 6, pp. 3432-3436, San Francisco, California, December 1-5,2003.

[10] A. Behzad and I. Rubin, "Power controlled multiple access control for wireless access nets," *in Proceedings of the 57th IEEE Semiannual Vehicular Technology Conference (VTC),* Vol. 1, pp. 423-427, Jeju, Korea, April 22-25, 2003.

[11] A. Behzad and I. Rubin, "High transmission power increases the capacity of adhoc wireless networks," in *Proceedings of IEEE International Conference on Communications (ICC),* Paris, France, June 20-24, 2004.

[12] C. Toh, "A novel distributed routing protocol to support adhoc mobile computing", in *Proceedings of IEEE 15th Annual Int'l. Phoenix Conf. Comp. and Commun.,* 1996.

[13] I. Rubin X. Huang and Y. Liu, "A QoS oriented topological synthesis protocol for mobile backbone networks (MBNs)," in *Proceedings of IEEE Vehicular Technology Conference (VTC Fall 2003),* Orlando, Florida, October 2003.

[14] I. Rubin and Y. Liu, "Link stability models for QoS adhoc routing algorithms," in *Proceedings of IEEE Vehicular Technology Conference (VTC Fall 2003),* Orlando, Florida, October 2003.

[15] I. Rubin, R. Khalaf, A. Moshfegh, and A. Behzad, "Delay-throughput performance of load-adaptive power-controlled multi-hop wireless networks with scheduled transmissions," in *Proceedings of IEEE VTC,* Orlando, Florida, October 4-9, 2003.

[16] S. Benjamin and I. Rubin, "Connected disc covering and applications to mobile gateway placement in adhoc networks," in *Proceedings of ADHOC Networks and Wireless (ADHOC-NOW)*, Toronto, Canada, 2002.

Chapter 10

VERTEX-LINKED INFRASTRUCTURE FOR ADHOC NETWORKS

VICTOR ON-KWOK LI

Department of Electrical and Electronic Engineering, University of Hong Kong, Hong Kong, China

Abstract: An adhoc network is composed of geographically dispersed nodes that may move arbitrarily and communicate with each other without the support of a stationary infrastructure. Compared with a wireless network with a stationary infrastructure, such as a cellular network, an adhoc network is inherently less efficient. Therefore, a number of proposals have been made to develop a quasi-stationary infrastructure for adhoc networks. However, the dynamic nature of adhoc networks makes it very .costly to maintain such an infrastructure. This article proposes a Vertex-Linked Infrastructure (VLI) for adhoc networks. This novel approach uses an easily deployable, survivable, wired infrastructure as a backbone of the adhoc network, thus realizing the advantages of an infrastructure in wireless communications, but without the overhead due to maintaining such an infrastructure.

1.   INTRODUCTION

An adhoc network, also known as a multi-hop packet radio network, is composed of user nodes that may move arbitrarily and communicate with each other without the support of a stationary infrastructure. Research in such networks is initiated in the Defense Advanced Research Projects Agency (DARPA) packet radio network[1]. They may be used in emergency search-and-rescue operations, battle field operations and data acquisition in inhospitable terrains.

An adhoc network is inherently less efficient than a wireless network with an infrastructure, such as a cellular network. As there is no wired infrastructure, the relatively limited wireless bandwidth is used to find and maintain routes as well as to transmit data. As its size grows, the amount of information required to be transmitted and to be maintained by each node in an adhoc network grows exponentially. The problem is exacerbated by topological changes. As nodes may move randomly and organize themselves arbitrarily, the topology of the network may change rapidly and unpredictably. When the current route is unusable, a new one must be re-established. This requires the transmission of many update and control messages in the precious wireless channel. With a wired infrastructure, as mobile nodes roam around the service area and get affiliated with different backbone nodes, such control messages can be transmitted in the relatively less congested wired channels. In addition, the two-level hierarchy of a backbone network, consisting of the backbone nodes, and the local access networks, each consisting of a backbone node and its affiliated mobile users, reduces the number of update messages required in the system. This is because the update messages only need to be propagated in the backbone network, and not in each individual local access network. In Chang and Li[2], a performance comparison is made between a packet radio network with such a two-level hierarchical structure and one which is fully distributed, and it is found that in terms of end-to-end throughput, the hierarchical structure outperforms the fully distributed network in most scenarios.

This is perhaps why, when there is a choice, as in most existing commercial systems, an infrastructure is used. Thus a cellular network has an infrastructure in the form of a wired network of stationary base stations, a satellite network, in the form of stationary ground stations, and a WiFi network, in the form of stationary access points.

In some applications, such as in a battlefield, it has traditionally been thought that a stationary infrastructure is impractical. After all, the fixed base stations will be easy targets for the enemy. However, even in such cases, because of the inherent advantages of having an infrastructure, there are various proposals for quasi-stationary infrastructures. The major difficulty with such approaches, however, is with the maintenance of the infrastructure. Therefore, it is desirable to have an adhoc network design with a stationary infrastructure, thus eliminating the overhead due to infrastructure maintenance, and yet is practical and survivable in a hostile environment such as a battlefield. In this paper, we propose the concept of a stationary, wired infrastructure for an adhoc network, called the Vertex-Linked Infrastructure (VLI). An adhoc network operating with a VLI is called a VLINET.

The rest of this paper is organized as follows. In Section 2, we describe some of the existing quasi-stationary infrastructures employed in adhoc networks. In Section 3, we introduce VLINET. Details on network operations, and topological and reliability considerations are included. In Section 4, we discuss ways to enhance the survivability of this proposed system, and consider optical fiber as a possible transmission medium. We conclude in Section 5.

2. QUASI-STATIONARY INFRASTRUCTURES

Although it is generally believed that a stationary infrastructure is impractical in a hostile environment, due to the inherent advantages of having an infrastructure, there are various proposals for adhoc networks with quasi-stationary infrastructures.

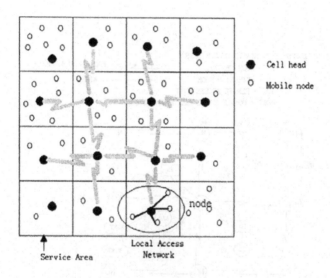

Figure -1. Cellular Packet Radio Network

For example, in the US High Frequency Intra-Task Force (HF-ITF) Network[3], the nodes in the network are organized into a set of clusters, each with a cluster head, and connected by a backbone network constituted from gateways and cluster heads. A distributed algorithm, called the Link Cluster Algorithm, provides the construction and maintenance of this two-level hierarchical organization. A critical assumption of this algorithm, which uses the Time Division Multiple Access (TDMA) technique to transmit control messages, is that each node must know the number of nodes in the network. This assumption may not hold in many applications. In Chang and Li[4], a Distributed Cellular Packet Radio Network (DCPRNET) is proposed. As shown in Fig. 1, the whole service area is divided into disjoint regions called cells. Each cell has a node elected as the cell head, which provides local network control functions, such as routing and flow control, to the nodes within the same cell. Each node is assumed to have Global Positioning System (GPS) capability, and with a map of the network layout, it knows which cell it is affiliated with. In other words, nodes are organized into clusters (local access networks) based on their geographical locations. The nodes will communicate with each other through the backbone network formed by the cell heads.

In Pond and Li[5,6], a hierarchical architecture for a distributed media access protocol is developed for the US Army's Enhanced Position Location and Reporting System (EPLRS)[7]. Again, the goal is to capture the advantage of reduced overhead available with a quasi-stationary infrastructure.

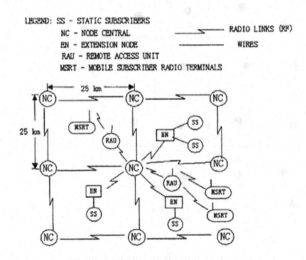

Figure -2. The US Army Mobile Subscriber Equipment (MSE) Network.

The US Army Mobile Subscriber Equipment (MSE) Network[8] , is another example. As shown in Fig. 2,MSE consists of an infrastructure with Node Centrals (NCs), Extension Nodes (ENs), and Remote Access Units (RAUs). All network elements are packaged on mobile platforms, but once deployed, will remain stationary. The EN's serve static subscribers, while the RAUs serve Mobile Subscriber Radio Terminals (MSRTs). An MSRT accesses its RAU by a radio link, from up to 15 km away. Each NC is typically connected by line-of-sight radios to four other NCs, with all the NCs located on a grid pattern with 25 km spacing. Each MSRT will be affiliated with an NC through its RAU. Communications between two MSRTs will go over the backbone network formed by the NCs.

More recently, there have been a number of proposals to deploy a quasi-stationary backbone network in the form of Unmanned Airborne Vehicles (UAVs) above the service area. Each UAV serves as the cluster head of some nodes within its coverage area. The UAVs communicate with each other with radio links. Due to the vulnerability of the UAVs to enemy attacks, and the mobility of the mobile users, this quasi-stationary backbone infrastructure is again very costly to maintain.

3. VERTEX-LINKED INFRASTRUCTURE

Given the advantages available with an infrastructure, it is natural to ask if it is possible to come up with an adhoc network design with a wired infrastructure, and yet is practical and survivable in a hostile environment such as a battlefield. We believe the answer is "yes." In this section, we propose the concept of a stationary, wired infrastructure, called the Vertex-Linked Infrastructure (VLI). An adhoc network operating with a VLI is called a VLINET.

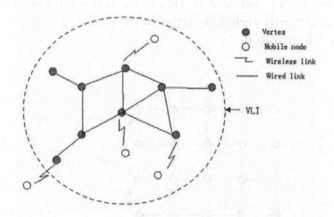

Figure -3. The Vertex-Linked Infrastructure (VLI) Network

As shown in Fig. 3, a VLI consists of a collection of transceivers connected by wires in an arbitrary topology. In this paper, to distinguish between these transceivers, which are stationary once deployed, and the mobile users, we call the former vertices and the latter nodes. A VLI will be deployed, perhaps from the air, in the service area of interest. Mobile nodes communicate with each other through this infrastructure, in much the same way as in a cellular system. The following describes the details of the operations of a VLINET.

3.1   Network Initialization and Deployment

Since a VLI has a static topology, one can pre-determine the paths between any two vertices (transceivers) in VLI before deployment. The pre-determined paths from one vertex to all other vertices can be stored locally at each vertex. For example, we can employ a self-routing address scheme whereby the path to each destination is encoded in the destination address. For survivability, multiple addresses, each corresponding to a different path, can be encoded for each destination.   More details are given in the next section on self-routing address design. Each vertex periodically broadcasts a beacon with its identity (ID). Each mobile user must first register with one of the vertices. If it receives beacons from multiple vertices, it can just pick one of them to register.   Again, for survivability, we can allow a mobile to register at multiple vertices. Using the wired network, the vertices periodically send updates on mobile users registered locally to other vertices in the VLI.   Thus each vertex has a complete picture of the vertex affiliation of each mobile user. Consider a mobile user A attempting to send to mobile user B.   A will send the packet with B's ID to A's affiliated vertex in the VLI. This vertex looks up the affiliated vertex of B, put this vertex's address into the packet header, and send it. Due to the self-routing nature of the address, the packet will eventually arrive at the affiliated vertex of B, and subsequently delivered to B.

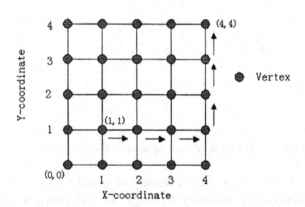

Figure -4. A four by four grid network.

3.2   Self-routing Address Design

Various self-routing address designs have been proposed for networks with a regular structure, such as the Grid Network (GN). As shown in Fig. 4, in the GN, the vertices are laid out in a grid pattern, and each has an address equal to its (x,y) coordinates. When a packet is routed in the network, a comparison is made between the x-coordinate of the destination address and that of the local address. Depending on whether it is smaller or larger than the local address, the packet is transmitted to the next node to the west or to the east. If the x-coordinates are the same, then we compare the y-coordinates to see whether we should go north (when the destination y-coordinate is larger than the local y-coordinate) or go south (when the destination y-coordinate is smaller than the local y-coordinate). If both the x- and y-coordinates are the same, then we know we have arrived at the destination. Suppose we are at vertex (1,1) and the destination is (4,4). By doing these x- and y-coordinate comparisons at all the intermediate vertices, we will take the packet three hops east along the x-coordinate, and then three hops north along the y-coordinate. Note that if we randomize the order of comparisons, i.e., instead of always comparing the x-coordinates first, we may compare the y-coordinates first in some cases, we will be able to obtain all the possible paths between (1,1) and (4,4). A similar scheme can be developed for other topologies with regular structures, such as the Shufflenet, or a hypercube network. Recently, a self-routing scheme has also been developed for a network with an arbitrary topology[9].

4.   DISCUSSION

In this section, we explore various enhancements of the basic VLINET described above, including flood search schemes, topological considerations, survivability considerations, and the choice of the transmission medium.

4.1   Additional Routing Schemes

The basic routing scheme for VLI as described in the last section relies on the self-routing address of vertices in the VLI. Thus, once the vertex with which the destination is affiliated is found, the packet will be routed to this vertex automatically. To enhance reliability, multiple alternative paths may be used. For an VLI with a regular topology, such as the Grid Network described in the last section, multiple paths may be obtained by randomizing the order of comparison of the x- and y-coordinates as a packet is forwarded. For an VLI with an arbitrary topology, the scheme described in Yuan et al.[9]

may be used. In this scheme, since each address corresponds to a fixed path, multiple redundant addresses, each encoding a different path, will be required. In the unlikely event none of these pre-selected redundant paths is available, due to excessive losses of vertices and links, it is still possible to transmit to the destination by a flooding scheme. The nice property about a flood search scheme is that if a path exists from the source to the destination, it will be found. The price to pay is the large number of redundant flood search messages. For example, the basic flooding scheme employed by MSE, or some of the modified flood search schemes described in Li and Chang[10] may be used. A flood search scheme will of course only be used as a last resort.

4.2 Topological consideration

As is described in the last section, a VLI does not have to conform to a specific topology. It may be a regular topology such as a grid, a star, or a ring, or it may be any arbitrary topology which may be most suited to the service area of interest. In fact, a customized topology may be designed to suit the terrain of the service area. For survivability considerations, it is probably best to consider those topologies with multiple redundant paths between pairs of vertices.

4.3 Survivability considerations

For survivability, and for improved data transport capacity, it is possible to deploy multiple VLINETs in the same service area. When a particular region of the service area lacks coverage, due to some vertices or links being destroyed, or due to increased data transmissions, one can rapidly deploy additional VLINETs. The network operates in pretty much the same way as described in Section 3. The only difference is when the origin and destination nodes are affiliated with different VLINETs, and there must be some way for us to bridge different VLINETs. One possibility is to allow vertices to communicate with each other over the wireless channel. Presently, they communicate with each other through the wired channels, and the more precious wireless resources are reserved for mobile users. An alternative is to deploy wireless repeaters in the system, whose sole purpose is to bridge multiple VLINETs. Thus selected vertices in each VLI will be employed as gateways for communications with other VLIs. We are effectively introducing an additional layer of hierarchy, consisting of the gateway nodes, in the system. As described in Yuan et al.[9], the self-routing address scheme for arbitrary topology can be easily extended to multiple hierarchies.

4.4 Transmission medium

Another important consideration is the choice of transmission media in the VLINETs. We believe a fiber-based infrastructure will be desirable. Optical communications is less susceptible to interference such as Electromagnetic Pulse (EMP) generated by the enemy. It is also possible to have all-optical infrastructure with fiber links and all-optical switches. The major limitation in optical communications is the limited optical logic processing capability. Fortunately, the self-routing address scheme developed in Yuan et al.[9] requires only simple single-bit optical processing and can be readily implemented with existing optical logic. In addition, with the development of radio on fiber technologies, it is possible to distribute the transceivers geographically within a service area. In Li et al.[11], we have developed RaFiNet, a radio-over-fiber implementation of a VLINET.

5. CONCLUSIONS

Compared with a wireless network with a stationary infrastructure, such as a cellular network, an adhoc network is inherently less efficient. Therefore, a number of proposals have been made to develop a quasi-stationary infrastructure for adhoc networks. However, the dynamic nature of adhoc networks makes it very costly to maintain such an infrastructure. This article proposes a Vertex-Linked Infrastructure (VLI) for adhoc networks. This novel approach uses an easily deployable, survivable, wired infrastructure as a backbone of the adhoc network, thus realizing the advantages of an infrastructure, but without the overhead due to maintaining such an infrastructure.

ACKNOWLEDGEMENT

This research is supported in part by the Area of Excellence Scheme established by the University Grants Committee of the Hong Kong Special Administrative Region, China (Project No. AoE/E-01/99).

REFERENCES

1. J. Jubin, J. Tornow, "The DAPRA packet radio network protocols," *Proc. IEEE,* Vol. 75, No. 1,1987, pp. 21-32

2. R.F. Chang and V.O.K. Li, "Comparison of hierarchical and fully distributed mobile packet radio network," *Proc. IEEE TENCON,* Hong Kong, September 1990, pp. 63 - 67.

3. J. Baker and A. Ephremides, "The architectural organization of a mobile radio network via a distributed algorithm," *IEEE Trans. on Comm.,* Vol. COM-29, No. 11, Nov. 1981, pp. 1694-1014.

4. R.F. Chang and V.O.K. Li, "Hierarchical routing in mobile packet radio networks," *Proc. IEEE Singapore International Conference on Networks,* Singapore, July 1989, pp. 427-432.

5. L. Pond and V.O.K. Li, "A distributed media access protocol for packet radio networks and performance analysis, Part I: network capacity," *International Journal of Communication Systems,* Volume 8, No. 1, Jan-Feb 1995, pp. 27 - 48.

6. L. Pond and V.O.K. Li, "A distributed media access protocol for packet radio networks and performance analysis, Part II: network setup time and data rate," *International Journal of Communication Systems,* Volume 8, No. 1, Jan-Feb 1995, pp. 49 - 64.

7. J.A. Kivett and R.E. Cook, "Enhancing PLRS with user-to-user data capability," *IEEE PLANS: Position, Location, and Navigation Symposium,* 1986, pp. 154– 161.

8. D. Schaum et al., "MSE Mobile Subscriber Equipment," *Army Communication,* Vol. 9, No. 3, Fall 1984, pp. 6 – 22.

9. X. C. Yuan, V.O.K. Li., C.Y. Li, and P. K. A. Wai, "A novel self-routing address scheme for all-optical packet switched networks with arbitrary topology," *IEEE/OSA Journal of Lightwave Technology,* Vol. 21, No. 2, Feb 2003, pp. 329 - 339.

10. V.O.K. Li and R.F. Chang, "Proposed routing algorithms for the U.S. Army Mobile Subscriber Equipment (MSE) Network," *Proc. IEEE MILCOM,* Monterey, California, October 1986, pp. 39.4.1-39.4.7.

11. V.O.K. Li et al., "RaFiNet: An implementation of a Vertex-Linked Infrastructure for adhoc networks," in preparation.

Chapter 11

PROBABILISTIC METHODS FOR LOCATION ESTIMATION IN WIRELESS NETWORKS

PETRI KONTKANEN, PETRI MYLLYMÄKI, TEEMU ROOS, HENRY TIRRI, KIMMO VALTONEN, HANNES WETTIG
Complex Systems Computation Group, Helsinki Institute for Information Technology, University of Helsinki & Helsinki University of Technology, P.O.Box 9800, 02015 HUT, Finland

Abstract: Probabilistic modeling techniques offer a unifying theoretical framework for solving the problems encountered when developing location-aware and location-sensitive applications in wireless radio networks. In this paper we demonstrate the usefulness of the probabilistic modelling framework in solving not only the actual location estimation (positioning) problem, but also many related problems involving pragmatically important issues like calibration, active learning, error estimation and tracking with history. Some interesting links between positioning research done in the area of robotics and in the area of wireless radio networks are also discussed.

1. INTRODUCTION

The location of a mobile terminal can be estimated using radio signals transmitted or received by the terminal. The problem is called with various names such as location estimation, geolocation, location identification, location determination, localization, and positioning. The traditional, geometric approach to location estimation is based on angle and distance estimates from which a location estimate is deduced using standard geometry. Instead of the geometric approach, we consider the probabilistic approach which is based on probabilistic models that describe the dependency of observed signal properties on the location of the terminal, and the motion of the terminal. The models are used to estimate the terminal's location when signal measurements are available.

The feasibility of the probabilistic approach in the context of wireless networks has already been demonstrated to some extent in a number of recent papers (Castro et al., 2001; Ladd et al., 2002, Roos et al., 2002a, Roos et al., 2002b; Schwaighofer et al., 2003; Youssef et al., 2003). Probabilistic methods have also been extensively used in robotics where they provide a natural way to handle uncertainty and errors in sensor data (Smith et al., 1990; Burgard et al., 1996; Thrun, 2000). In a recent survey (Thrun, 2003), Sebastian Thrun summarizes the central role of probabilistic methods in robotic mapping as follows:

"Virtually all state-of-the-art algorithms for robotic mapping in the literature have one common feature: They are probabilistic. [...] The reason for the popularity of probabilistic techniques stems from the fact that robot mapping is characterized by uncertainty and sensor noise. "

Many of the probabilistic methods developed in the robotics community, in particular those related to mapping, location estimation and tracking, are also applicable in the context of wireless networks. In the following we discuss selected topics in probabilistic location estimation, many of which are well-known in probabilistic modeling, but have received relatively little attention in the domain of wireless networks.

We focus primarily on wireless local area networks, WLANs, but most of the ideas and concepts are applicable to many other wireless networks as well, including those based on GSM/GPRS, CDMA or UMTS standards. The rest of the paper is organized as follows: In Section 2 we discuss *calibration,* the process of obtaining a model of the signal properties at various locations. The actual location estimation and tracking phase following calibration is considered in Sections 3 and 6. Issues related to the optimal choice of calibration measurements are discussed in Section 4. In many cases, it is useful to complement a location estimate with information on its accuracy; in Section 5 we describe methods for error estimation and visualization. Conclusions are summarized in Section 7.

2.   CALIBRATION

In order to obtain a positioning model, we need to estimate the distribution of the signal properties, e.g., signal strength, as measured by the device to be localized for the various locations in consideration. This has traditionally been done using knowledge of radiowave propagation. Several propagation prediction or *cell planning* tools are available for this purpose (Andersen et al., 1995; Wölfle and Landstorfer, 1999). We adopt an empirical approach, i.e. we estimate the required distributions from *calibration data* gathered at different locations in consideration. Experimental studies suggest that propagation methods are not competitive against empirical models in terms of positioning accuracy due to insufficiently precise signal models (Bahl and Padmanabhan, 2000, Roos et al., 2002b).

Consider a finite set of calibration points l, which are labelled by their x and y coordinates (and possibly a third coordinate z or other additional information). For each calibration point we gather calibration data, i.e., a number of observation vectors o to estimate the distribution of signal properties from. For a discussion on how this can and should be done for a given set of calibration points and observations see (Castro et al., 2001; Roos et al., 2002a; Schwaighofer et al., 2003; Youssef et al., 2003). When we then want to position a device with current signal readings o, we calculate the probabilities $p(l \mid o)$ for each possible location l using the Bayes rule and the distributions estimated from the calibration data as described in Section 3. For simplicity, we assume that the set of possible locations can be considered equal to the set of calibration points. If this is not the case, e.g., when a continuous location variable is used, we need to interpolate in order to obtain a distribution of the signal properties at locations from which no calibration data is available. Note that without interpolation we can model only a finite—and for practical reasons preferably not too large—set of possible locations. We can then determine for example in which room a device is located (with certain probability), but have no probabilities associated to locations in between the calibration points.

But how should we choose the set of possible locations and how should we collect calibration data? A simple solution is to use a *probability grid* (Burgard, 1996), dividing the positioning space into cells of some size, e.g., 1m × 1m. In order to obtain a distribution of the signal properties at each grid point without interpolation, one then needs to collect a sufficient number of training vectors at each grid point. However, it may be impractical to remain at each grid point for the time it takes to gather enough data—and favorably move around in order to capture variance due to orientation and within the area of the cell—before moving on to the next cell.

A more convenient way is to gather data vectors continuously just walking (or driving) around. We only need to record the time label of each observation and the time labels and coordinates of those locations at which the calibrator changes direction and/or speed. This way we quickly obtain a large number of observations equipped with their exact location. However, we (usually) get only one observation per location, which does not suffice to reliably estimate the distribution of signal properties. Furthermore it is computationally problematic to deal with such a large number of possible locations in a model; note that when a device supplies us with an observation vector every 500ms, a calibration round of an hour already yields up to 7200 locations.

A natural way of dealing with this situation is to group the locations into clusters (Youssef, 2003). Each cluster should consist of a sufficient number of vectors to supply a good estimate of the signal properties in its area, and as its location we may take—for example—the center of gravity of its measurements' positions. An interesting and theoretically appealing way to produce such clustering is given by the principle of *Minimum Description Length* (MDL) in its most recent form, the *Normalized Maximum Likelihood* (NML) (Rissanen, 1996), for details see (Kontkanen et al., 2004). Figure 1 shows such clustering of a calibration tour. Note, that there is no need to decide on the number of clusters in advance, the algorithm will choose as many as can reliably be distinguished from the data collected.

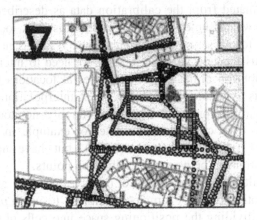

Figure 1 NML clustering of signal data collected continuously along a calibration tour. Each circle represents one vector of measurements gathered at its position, the different clusters are colour-coded.

3.   LOCATION ESTIMATION

After the calibration phase we have, for any given location l, a probability distribution $p(o \mid l)$ that assigns a probability (density) for each measured signal vector o. By application of the Bayes rule, we can then obtain the so called *posterior distribution* of the location (Roos et al., 2002b):

$$p(l \mid o) = p(o \mid l)\, p(l)\, /\, p(o) = p(o \mid l)\, p(l)\, /\, \Big(\textstyle\sum_{l' \in L} p(o \mid l')\, p(l')\Big),$$

where $p(l)$ is the *prior probability* of being at location l before knowing the value of the observation variable, and the summation goes over the set of possible location values, denoted by L. If the location variable is continuous, the sum is replaced by the corresponding integral.

The prior distribution $p(l)$ gives a principled way to incorporate background information such as personal user profiles and to implement tracking as described in Section 6. In case neither user profiles nor a history of measured signal properties allowing tracking are available, one can simply use a uniform prior which introduces no bias towards any particular location. As the denominator $p(o)$ does not depend on the location variable l, it can be treated as a normalizing constant whenever only relative probabilities or probability ratios are required.

The posterior distribution $p(l \mid o)$ can be used to choose an optimal estimator of the location based on whatever loss function is considered to express the desired behavior. For instance, the squared error penalizes large errors more than small ones, which is often useful. If the squared error is used, the estimator minimizing the expected loss is the expected value of the location variable:

$$E[l \mid o] = \textstyle\sum_{l \in L} l\, p(l \mid o),$$

assuming that the expectation of the location variable is well defined, i.e., the location variable is numerical. Location estimates, such as the expectation, are much more useful if they are complemented with some indication about their precision. We discuss error estimates in Section 5 below.

The presented probabilistic approach can be contrasted with the more traditional, geometric approach to location estimation used in methods such as angle-of-arrival (AOA), time-of-arrival (TOA), and time-difference-of-arrival (TDOA). In the geometric approach the signal measurements are transformed into angle and distance estimates from which a location estimate is deduced using standard geometry. One of the drawbacks of the geometric approach is that there is no principled way to deal with the incompatibility of the angle and distance estimates caused by measurement errors and noise. On the other hand, the geometric approach is usually computationally very efficient.

4.   ACTIVE LEARNING

In Section 2 we only considered the problem of obtaining a model of the signal properties given training data collected from known locations. The resulting model is strongly dependent on where and how much training data is collected. Obviously, the training data should not leave large areas uncovered or otherwise there would be no way to reliably infer the signal properties in such areas. Also, for various reasons, for some areas the signal model is required to be more accurate than in general, in order to achieve accurate location estimation. For instance, two distinct locations may be roughly similar in terms of signal properties so that they can be told apart only by a small margin. In such areas, more extensive calibration is required.

In practice, if it is possible to collect a large amount of training data, a reasonable calibration result is obtained by collecting training data roughly uniformly from each location. Areas where the signal properties are expected to vary within small distances due to, for instance, large obstacles, may be better covered with relatively higher density, whereas large open areas where the signal is likely to be constant, can be left with less attention. In case extensive calibration is costly or otherwise impossible, it becomes critical to choose the calibration points as well as possible. The problem of choosing optimal actions in order to reduce uncertainty has been studied in the robotics literature under the name *robotic exploration.* In general, optimal decision strategies are intractable and various heuristics are used (Burgard et al., 2000; Thrun, 2003).

A practical method for locating potentially useful candidates for new calibration points is based on the estimate of the future expected error. This estimate is calculated by summing over all possible future observations *o*:

$$E\,[err\,|\,l\,] = \sum_o E\,[err\,|\,l,o]\,p(o),$$

where *l* is the calibration point candidate, and $E[err\,|\,l,o]$ is the expected error:

$$E\,[err\,|\,l,o\,] = \sum_{l'\,\in\,L} p(l'\,|\,o)\,d(l',l),$$

for the preferred distance function *d.*

The candidate points *l* can be chosen by using a tight grid. For example, the grid spacing could be approximately one meter. One or more grid points with a high expected error, or points surrounded by several such grid points, are then used as new calibration points. If the dimensionality of the observation vector *o* is so high that the summing over all *o* as above is not feasible, the sum can be approximated by sampling. An ever simpler approach is to use the calibration data as the set over which the sampling is performed, in which case one only needs to sum over the calibrated observations.

To implement the method based on equations above, one needs to determine the probability distribution or density over the future observations. In practice, it has to be approximated from the calibration data. One possible approximation method is as follows. When computing E[*err* | *l*] for some location *l*, one replaces the *p(o)* by the probability distribution based on the past observations made at the calibration point closest to *l*. The efficiency of the method can then be further improved by approximating E[*err* | *l,o*] by *d(l*, l)*, where *l** is the point estimate produced by the positioning system after seeing observation *o*.

5.   ERROR ESTIMATION AND VISUALIZATION

In order to visualize the uncertainty associated with the location, we assume that we have a probability distribution *p*, either a probability mass function or a density, which describes the uncertainty about the actual location. In addition to reporting a point estimate—here taken to be the expected value—we can visualize the uncertainty related to distribution *p*. This can be done, for instance, by drawing an ellipse centered at the expected location such that the orientation and size of the ellipse describes the uncertainty of the location estimate as well as possible.

As a first step of obtaining such an "uncertainty ellipse" one first needs to obtain certain summary statistics from the distribution *p*. These statistics are, in addition to the expectation, contained in the *variance-covariance matrix*. The variance-covariance matrix describes the variance of the location in both *x* and *y* coordinates together with the correlation of the two coordinates. The second step is to evaluate the two *eigenvectors* of the variance-covariance matrix. This is a simple exercise in linear algebra. For instance, in case the two coordinates *x* and *y* happen to be independent in the distribution *p*, i.e., there is no correlation, the eigenvectors are parallel to the two coordinate axes. Finally, one displays an ellipse whose axes are parallel to those given by the two eigenvectors of the variance-covariance matrix. The lengths of the axes are given by the *eigenvalues* multiplied by a scaling constant. We give a rule for determining the value of the scaling constant below, after we have first discussed the interpretation of the ellipse.

One interpretation for the uncertainty ellipse is that assuming (pretending) that the estimated density of the location is bivariate Gaussian, the ellipse is the smallest area that contains a fixed probability mass. Given the probability mass to be covered by the ellipse, one can obtain the aforementioned scaling constant by taking the square root of the Chi-squared value with two degrees of freedom. For instance, if 95 % coverage is required, the scaling constant becomes $\sqrt{5.991} = 2.448$. An illustration of the error ellipse is shown in Fig. 2.

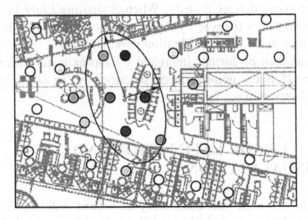

Figure 2 Uncertainty ellipse. Probabilities at a discrete set of locations are denoted by circles; dark shading implies high probability. The ellipse centered at the expected location has axes parallel to eigenvectors of the variance-covariance matrix and lengths proportional to eigenvalues.

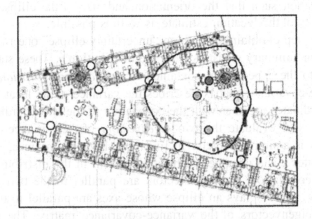

Figure 3 Uncertainty about the estimate represented by a polar coordinate system placed at the point estimate. Calibration points are marked by circles, colored depending on $p(x,y)$. The relative amount of uncertainty in each direction away from the point estimate is visualized by the curve.

Whereas the ellipse approach shows the uncertainty about location in two orthogonal directions with respect to the point estimate, a generalization to an arbitrary number of directions can be obtained by mapping $p(x,y)$ to a polar coordinate system centered on the point estimate. In this method, the origin is placed at the point estimate and each calibration point mapped to the polar coordinate system $a(x,y),d(x,y)$, where $a(x,y)$ is the angle w.r.t. the point estimate and $d(x,y)$ is the distance. It is convenient to discretize both $a(x,y)$ and $d(x,y)$, resulting in the case of two-dimensional space in a set of segments that partition the space disjointly and exhaustively.

We gain a discrete two-dimensional distribution $p_p(a(x,y), d(x,y))$ over the location space. The curve visualizing a wanted contiguous portion of the total mass can then be derived from $p_p(a(x,y), d(x,y))$. Relative distances from the origin are first determined for each sector based on expected distances. The resulting shape describes relative probability mass in each "direction" (sector). To represent the spread of uncertainty as well, the curve can be scaled so that it covers a desired fraction of $p_p(a(x,y),d(x,y))$. For a screen shot of an implementation, see Fig. 3.

6. TRACKING

Location estimation accuracy can be greatly improved if instead of a single signal measurement, a series of measurements is available unless the mobile device is moving with very high speed or the time interval between measurements is very long. Such a series of measurements allows keeping track of the device's location as a function of time, also called *tracking*. It is convenient to model the situation as a *hidden Markov model* (Rabiner, 1989) illustrated in Fig. 4.

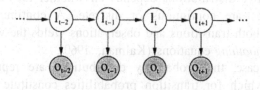

Figure 4 Hidden Markov model. State variables (white nodes) are hidden (not observed). Observed variables are denoted by shaded nodes. Horizontal arrows correspond to transition probabilities between successive states. Vertical arrows correspond to observation probabilities given state.

In a hidden Markov model, the variables $l_1, l_2 ...$ correspond to a sequence of states indexed by time t. In our location estimation domain, the state correspond to location and hence, the state sequence constitutes a trajectory of the located device. The model also has a set of corresponding observation variables, denoted by o_1, o_2 ... Each observation variable, o_t, is assumed to be dependent only on the current location, l_t. In the model in Fig. 4, the location at time t is dependent on the earlier locations only through the previous location l_{t-1}. Generalizations to higher order dependencies are easily expressed in the general framework of graphical probabilistic models (Cowell et al., 1999; Pearl, 1988).

The power of the hidden Markov model stems from the fact that inference in the model is effective. Given a series of observations, o_1, ..., o_n, the probability distribution of the location at any given time can be computed in order $O(n)$ operations using the standard probabilistic machinery developed for graphical models. Furthermore, maintaining the distribution of the current location, as observations are made one by one, can be done iteratively such that for each new observation, only constant, $O(1)$, time is needed. However, one should be cautious about the multiplicative factors hidden in the $O(n)$ and $O(1)$ notation. We return to this issue shortly below. Other possible inferences include tracking with a k step *lag*, i.e., maintaining the distribution of the location variable l_{t-k} instead of the most recent location, l_t. This is called *smoothing* as the evolution of the location variable l_{t-k} as a function of time t is smoother than the evolution of the current location l_t. The *Viterbi* algorithm gives the most likely trajectory given a sequence of observations, see (Rabiner, 1989).

In order to apply the hidden Markov model, one needs to specify two kinds of probabilities. First, one needs to determine the conditional probability distribution of the observation variable given the state variable. This is exactly the aim of calibration as discussed in Section 2. Second, the conditional distribution of each state s_t given the previous state s_{t-1}, called the *transition probability*, has to be determined. The form of these two kinds of conditional probability distributions depends on whether the location and observation variables are continuous or discrete. A continuous linear-Gaussian model for both transitions and observations yields the well-known *Kalman filter* and *smoothing* equations (Kalman, 1960).

In the discrete case, the probability distributions are represented as probability tables, which for transition probabilities constitute *an $N \times N$* matrix where N equals the number of possible locations. In the general case, the multiplicative factor in the $O(n)$ and $O(1)$ notation above for the computational complexity of inference is at least as large as N^2. Methods to reduce the computational complexity of tracking and smoothing when using discrete-valued location include the aforementioned clustering approach that reduces the number of locations N. In addition, a large proportion of state transition probabilities are usually extremely small or zero. In such a case the transition probability matrix is sparse which can be exploited to essentially reduce computational complexity. One can also resort to approximative inference using, for instance, *particle filtering* techniques that try to focus computation on areas of the state space where most of the probability mass lies (Fox et al., 1999).

7. CONCLUSIONS

We showed how the probabilistic modelling approach can be used for defining a unifying framework offering a theoretically solid solution to the location estimation problem, and what is more, also to many related, practically important problems involving issues like calibration, active learning, error estimation and tracking with history. Nevertheless, having said that, it must be acknowledged that problems in the real world are always more complicated than the textbook examples, and developing these theoretically elegant solutions to a robust, off-the-shelf software package like for example the Ekahau Positioning Engine (see www.ekahau.com), requires several minor but practically important technical tricks the details of which are outside the scope of this paper. However, we strongly believe that the best way to develop location-aware applications is to start with a theoretically correct, "ideal" solution, and then approximate that solution as accurately as possible given the pragmatic constraints defined by the real-world environment. Our experiences suggest that although it is not the only possible approach for this, the probabilistic modeling framework offers a viable solution for developing practical applications in this domain.

ACKNOWLEDGEMENTS

This work was supported in part by the IST Programme of the European Community, under the PASCAL Network of Excellence, IST-2002-506778. This publication only reflects the authors' views.

REFERENCES

Andersen, J. B., Rappaport, T. S., and Yoshida, S., 1995, Propagation measurements and models for wireless communications channels, IEEE Communications Magazine, 33:42–49.

Bahl, P., and Padmanabhan, V. N., 2000, Radar: An In-building RF-based user location and tracking system, in: Proc. 19th Annual Joint Conf. of the IEEE Computer and Communications Societies (INFOCOM-2000), Vol. 2, Tel-Aviv, Israel, pp. 775–784.

Burgard, W., Fox, D., Hennig, D., and Schmidt, T., 1996, Estimating the absolute position of a mobile robot using position probability grids, in: Proc. 13th National Conf. on Artificial Intelligence (AAAI-1996), Portland.

Burgard, W., Moors, M., Fox, D., Simmons, R., and Thrun, S., 2000, Collaborative multi-robot exploration, in: Proc. IEEE Int. Conf. on Robotics & Automation (ICRA-2000), San Francisco.

Castro, P., Chiu, P., Kremenek, T., and Muntz, R., 2001, A Probabilistic room location service for wireless networked environments, in: Proc. 3rd Int. Conf. on Ubiquitous Computing (UBICOMP-2001), Atlanta.

Cowell, R., Dawid, P. A., Lauritzen, S., and Spiegelhalter, D., 1999, Probabilistic networks and expert systems. Springer-Verlag, New York.

Fox, D., Burgard, W., Dellaert, F., and Thrun, S., 1999, Monte Carlo localization: Efficient position estimation for mobile robots, in: Proc. 16th National Conf. on Artificial Intelligence (AAAI-1999), Orlando, pp. 343–349.

Kalman, R. E., 1960, A New approach to linear filtering and prediction problems, Transactions of the ASME–Journal of Basic Engineering, 82(Series D):35–5.

Kontkanen, P., Myllymäki, P., Buntine, W., Rissanen, J., and Tirri, H., 2004, An MDL framework for data clustering, in: Advances in Minimum Description Length: Theory and Applications, P. Grünwald, I. J. Myung, and M. Pitt, eds., MIT Press.

Ladd, A. M., Bekris, K., Rudys, A., Kavraki, L. E., Wallach, D. S., and Marceau, G., 2002, Robotics-based location sensing using wireless Ethernet, in: Proc. 8th Annual Int. Conf. on Mobile Computing and Networking (MOBICOM-2002), Atlanta, pp. 227–238.

Pearl, J., 1988, Probabilistic Reasoning in Intelligent Systems: Networks of Plausible Inference. Morgan Kaufmann Publishers, San Mateo.

Rabiner, L. R., 1989, A Tutorial on hidden Markov models and selected applications in speech recognition, Proc. of the IEEE, 77(2):257–286.

Rissanen, J., 1996, Fisher information and stochastic complexity, IEEE Transactions on Information Theory, 42(1):40–47.

Roos, T., Myllymäki, P., and Tirri, H., 2002a, A Statistical modeling approach to location estimation, IEEE Transactions on Mobile Computing, 1(1):59–69.

Roos, T., Myllymäki, P., Tirri, H., Misikangas, P., and Sievänen, J., 2002b, A Probabilistic approach to WLAN user location estimation, Int. Journal of Wireless Information Networks, 9(3): 155–164.

Schwaighofer, A., Grigoras, M., Tresp, V., and Hoffmann, C., 2004, GPPS: A Gaussian process positioning system for cellular networks, in: 17th Annual Conf. on Neural Information Processing Systems (NIPS-2003), Vancouver.

Smith, R., Self, M., and Cheeseman, P., 1990, Estimating uncertain spatial relationships in robotics, in: Autonomous Robot Vehicles, Springer-Verlag, Berlin-Heidelberg, pp. 167–193.

Thrun, S., 2000, Probabilistic algorithms in robotics, AI Magazine, 21(4):93–109.

Thrun, S., 2003, Robotic mapping: A Survey, in: Exploring Artificial Intelligence in the New Millennium, Morgan Kaufmann Publishers, San Francisco, pp. 1–35.

Wölfle, G., and Landstorfer, F. M., 1999, Prediction of the field strength inside buildings with empirical, neural, and ray-optical prediction models, in: 7th COST-259 MCM-Meeting in Thessaloniki, Greece.

Youssef, M. A., Agrawala, A., and Shankar, A. U., 2003, WLAN Location determination via clustering and probability distributions, in: IEEE Int. Conf. on Pervasive Computing and Communications (PERCOM-2003), Fort Worth.

PART III

MOBILE WIRELESS INTERNET AND SATELLITE APPLICATIONS

Chapter 12

COPING WITH UNCERTAINTY IN MOBILE WIRELESS NETWORKS

SAJAL K DAS[1], CHRISTOPHER ROSE[2]

[1]*Center for Research in Wireless Mobility and Networking (CReWMaN), Department of Computer Science and Engineering, The University of Texas at Arlington, Arlington, TX 76019-0015, USA, E-mail: das@cse.uta.edu;* [2]*WINLAB, Dept of Electrical Engineering, Rutgers University, Piscataway, NJ 08854, USA, E-mail: crose@winlab.rutgers.edu*

Abstract: With the availability of inexpensive wireless devices including sensors, and ongoing emphasis on greater integration of components, extremely large-scale wireless networks of interconnected mobile devices are inevitable in near future. It is also envisioned that almost all these varied devices will require some form of Internet access. Unfortunately, the uncertainty associated with wireless mobile networks produces unique challenges to achieving seamless integration with the Internet while provisioning end-to-end quality of service (QoS). In particular, the uncertainty in wireless channels as well as node mobility, and hence network topology, can bedevil protocols more suited to a "classical" Internet structure. Therefore, new protocols have to be designed that must be (i) robust against the uncertainty in traffic load, host mobility, resource availability and wireless link characteristics; (ii) adaptive to the network dynamics, thus making learning and prediction integral components in the design methodology; and (iii) intrinsically on-line so as to make real-time decisions based on temporal and spatial information.

In order to cope with uncertainty in wireless mobile networks, we propose an overarching theoretical framework for representing relevant network information in terms of underlying entropies, entropy rates and their inter-relationships. We will demonstrate how to apply information theoretic learning and prediction tools for collection and dissemination of network state information that can be used for robust and adaptive protocol design. Specifically, we will investigate the applicability of this novel framework in designing optimal mobility tracking and resource management, and also coping with uncertainty in traffic load, topology control and routing.

1.   INTRODUCTION AND MOTIVATION

Recent years have witnessed an explosion in public demand for wireless access to the Internet. Furthermore, with greater integration of components, large scale deployment of inexpensive wireless sensor networks is also on the horizon. Therefore, extremely large increases in the number of interconnected mobile wireless devices are inevitable, and almost certainly all these devices will require some form of Internet access.

Unfortunately, there are unique problems associated with seamless integration of wireless systems into the Internet. We claim that a primary difference between wired and wireless systems is **uncertainty** which first manifests itself at the physical layer as rapidly varying link qualities or availabilities owing to both channel conditions (stochastic fluctuations in the bit-rate and bit-error rate of wireless links) and variability in network topology owing to mobile nodes. This variability wreaks havoc with standard Internet protocols owing to implicit underlying structural assumptions, often called *layered abstractions.*

For instance, it is well known that a delayed response from an endpoint in a wired network suggests congestion along some part of a route. Thus, since congestion is something caused by convergence of traffic at some point, protocols such as TCP/IP rightly require sources to immediately and strongly limit transmission (back-off) when congestion is detected. However, in wireless systems, the lack of an acknowledgment could occur for any number of reasons which are only somewhat dependent on how traffic is directed. Therefore, coping mechanisms such as back-off can be both ineffective and wrongheaded when dealing with the transient link issues associated with mobile wireless systems.

Furthermore, what begins as uncertainty associated with the physical layer of the system, percolates up through every layer of the protocol stack since each layer depends on the efficacy of methods used by the layer below. For example, there exists uncertainty in node mobility, wireless network topology and load, traffic characteristics, routing and resource (e.g., bandwidth) availability. The manner through which such percolation occurs can be extremely complex. Therefore, **quantifying, coping** with and/or **reducing** uncertainties are key challenges and perhaps *organizing principles* for understanding the roles of wireless network and mobility and their seamless and successful integration into the Internet.

Uncertainty is naturally **quantified** using information theoretic definitions of *entropy* [24] and *entropy rates*. Furthermore, a vast amount of technical machinery has been developed over the years [7] to deal with uncertainty in communications systems, and perhaps most importantly, to provide quantitative bounds on the input output relationships of *any* process which operates on information to produce some output. This should prove especially helpful in that it allows abstractions which do not necessarily have to consider the details of a given protocol or a specific model other than the effect on the underlying uncertainty.

One may **cope** with uncertainty by simply reacting to conditions as warranted. We define the term "reacting" broadly enough to include prediction of future values of parameters (such as channel state or node positions) and subsequent related action. To this end, research is needed to articulate appropriate responses to various wireless/mobility-induced network events under various scenarios as well as to understand how relevant information can be gathered and disseminated to appropriate places. For instance, if node positions are known or predictable, then timely knowledge of node *itineraries* would certainly help in efficient resource allocation. Whether it is possible to construct and maintain such itineraries, settling upon the proper levels of abstraction, and the necessary amount/degree of information dissemination are important research challenges. Similarly, for uncertainty due to the wireless channel, developing models which capture and reify the relevant features (such as fading channel or interference environment) from a network perspective requires similar consideration of appropriate information collection, abstraction and dissemination at the physical layer and above.

As opposed to simply coping with uncertainty, one might also ask which aspects of wireless system design are most deleterious to efficient network operation and seek to **reduce** them in some way. For instance, suppose the most pernicious aspect of wireless is physical channel variability, and the machinations necessary to overcome it using network protocols are shown to be prohibitive. If true, such a result suggests that effort would be best expended in controlling variability in the wireless channel using, for instance, recent advances in multiple antenna systems to better guarantee channel quality, or even more recent advances in exploiting mobility and channel variability for improved throughput. In a similar vein, one could imagine that with proliferation of various wireless devices, mutual interference is a serious problem. One could then imagine developing network services which helped wireless transmitters and receivers better coordinate their use of shared wireless resources, thereby controlling the interference environment to the extent possible.

In summary, we seek to **tame** uncertainty in wireless mobile networks by

- Developing an information theoretic framework for representing relevant network information in terms of underlying entropies, entropy rates and their inter relationships.

- Applying information theoretic learning and prediction tools for collection and dissemination of network state information that can be used for robust and adaptive protocol design.

- Developing design methodologies and simulation tools which will strongly impact protocol design.

The uniqueness of our approach lies in a unified treatment which applies known and powerful tools for dealing with uncertainty to a domain where they have not been previously applied with rigor or breadth.

2. WHY AN INFORMATION THEORETIC FRAMEWORK?

We believe that any credible analytic framework to deal with uncertainties must be based on *information theory* since it allows precise quantification and comparisons of what is known or not known. But perhaps even more important, it also provides constraints on what is *knowable* when the sets of stochastic processes which comprise any network are processed. Thus, information theory plays a pivotal role everywhere in network design, from the traditional application of methods for compact data transfer, to predicting future events based on past observations.

For example, consider the amount of information that must be disseminated through a dynamically changing network topology due to node mobility in order to do routing at some level of efficacy and efficiency. If node mobility is characterized by some set of random processes, then node position uncertainty can be characterized by an *entropy rate*. Node mobility causes topology changes in the connection graph which can again be quantified with an entropy rate. Finally, the topology changes cause changes in routing tables and again, routing table changes can be quantified in terms of entropy rates. The transformations from node position uncertainty to routing table uncertainty are complex, but at each stage, the amount of uncertainty is bounded by such simple results as the *data processing theorem* [7]. Thus, practical applications based on a quantitative understanding of uncertainty might include protocols which predicted what network information would be needed where, and delivered it in a timely fashion. Alternatively, an application might be as simple as an assessment of whether careful mobility management and associated routing should be pursued or abandoned.

This basic approach to uncertainty can be extended to other network parameters such as wireless link quality, traffic load, or resource demands. Thus, the framework is general and perhaps most importantly, the quantities involved can be measured and then characterized using standard information theoretic tools such as entropy maximization, the *MaxEnt* principle [10], the principle of insufficient reasons [6], the cross-entropy minimization principle [13] or the competitive analysis [5], to mention a few. Then once network uncertainty is quantified, one can also consider facilities/services to reduce uncertainty where needed.

3. COPING WITH UNCERTAINTY

If network processes were complex but not stochastic, then the existing information theoretic model-building methods might be able to provide perfect predictions of network parameters like node positions. Of course, even so there would still be difficult provisioning problems in wireless mobile networks associated with scheduling of resources in response to implicit network changes. However, the fact is that such perfect foreknowledge is impossible to obtain, and this lack of certainty adds to the problem difficulty in at least two ways. First, purely reactive provisioning is difficult and second, providing the necessary location information where it is needed throughout the network requires network resources that could otherwise be used for other traffic. Our goal is to quantify this uncertainty and determine bounds on how well it can be reduced and/or controlled throughout the network.

3.1 Mobility Tracking

We begin by reviewing a successful application of **lossless coding** to mobility model characterization since many of the basic ideas will be used throughout this paper. It was first shown by Bhattacharya and Das [2, 3] that it is impossible to optimally track a mobile node with less information exchange between the network and the mobile than the entropy rate (bits/second) of the mobility process. Specifically, given all past observations of node position and the best possible predictors of future node position, some uncertainty in node position will *always* exist unless the node and the network exchange location information. The *method* by which this exchange takes place (some combination of paging and registration) is irrelevant to this bound. All that matters is that the exchange exceeds the mobility process entropy rate.

So, a key issue in establishing bounds is characterization of the mobility process (and therefore its entropy rate) in an adaptive manner. To this end, based on the information theoretic framework, an optimal on-line adaptive location management algorithm/protocol, called **LeZi-update,** was designed in [2, 3] for cellular personal communication networks. Rather than assuming a standard mobility model of the node, LeZi-update learns node movement history stored in a Lempel-Ziv (LZ) type of compressed dictionary, builds a **universal** mobility model by minimizing entropy, and predicts future locations with a high degree of accuracy. In other words, LeZi-update offers a model-independent solution to manage uncertainty related to node mobility. This generality is particularly useful for protocol design in wireless mobile networks where node characteristics are unknown or change with time.

As shown in Figure 1, the LeZi-update scheme uses a *symbolic space* to represent every cell of the wireless cellular network as an alphabetic symbol and captures the movement history of the mobile as a string of symbols. That is, while the geographic location data are often useful in obtaining precise location coordinates, the symbolic information removes the burden of frequent coordinate translation and is capable of achieving universality across different networks. Tacit in this formulation is that every node has some movement patterns which can be learned in an on-line fashion. Essentially, we assume that node itineraries are inherently *compressible* and this allows application of *universal data compression* algorithms [27] which make very basic and broad assumptions, and yet minimize the source entropy for stationary, ergodic, stochastic processes.

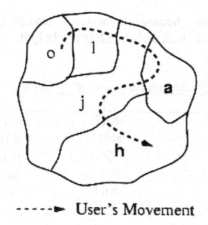

----→ User's Movement

Figure 1 Symbolic representation of mobility

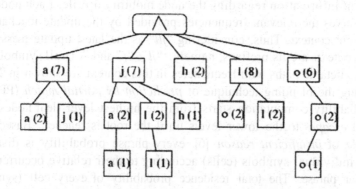

Figure 2 Trie holding cells and their frequencies

In LeZi-update, the symbols (cell-ids) are processed in chunks and the entire sequence of symbols withheld until the last update is reported in a compressed (encoded) form. For example, the input sequence

ajlloojhhaajlloojaajlloojaajll...

is parsed as distinct substrings (phrases) "*a, j, l, lo, o, jh, h, aa, jl, loo, ja, aj, ll, oo, jaa, jll, ...*". As shown in Figure 2, such a symbol-wise context model, based on variable to fixed-length coding, can be efficiently stored in a dictionary implemented by a *trie*. Essentially the mobile acts as an **encoder** while the network acts as a **decoder** and the frequency of every symbol is incremented for *every prefix of every suffix* of each phrase. By accumulating larger and larger contexts, one can effect a paradigm shift from traditional position update to *path update*. For stationary ergodic sources with *n* symbols, this framework achieves asymptotic optimality, with improved update cost bounded by $\Omega(\lg n - \lg \lg n)$ where lg *n* denotes logarithm base 2.

Table 1   Phrases and their frequencies at context "jl", "j" and Λ

jl	j	Λ		
l|jl(1)	a|j(1)	a(4)	aa(2)	aj(1)
Λ|jl(1)	aa|j(1)	j(2)	ja(1)	jaa(1)
	l|j(1)	jl(1)	jh(1)	l(4)
	ll|j(1)	lo(1)	loo(1)	ll(2)
	h|j(1)	o(4)	oo(2)	h(2)
	Λ|j(2)		Λ(1)	

One major objective of this update scheme is to endow the paging process, by which the system finds nodes whose position is uncertain, with sufficient information regarding the node mobility profile. Each node in the trie preserves the relevant frequencies provided by the update mechanism in the current context. Thus, considering *"jll"* as the latest update message, the usable contexts are its prefixes, namely: *"jl"*, *"j"* and Λ (null symbol). A list of all predictable paths with frequencies in this context are shown in Table 1. Following the blending technique of *prediction by partial match* (PPM) [6], the probability computation starts from the highest level (leaf nodes) of the trie and *escapes* to the lower levels until the root is reached. Based on the principle of *insufficient reason* [6], every phrase probability is distributed among individual symbols (cells) according to their relative occurrence in a particular phrase. The total residence probability of every cell (symbol) is computed by adding all the probabilities it has accumulated from all possible phrases at this context. The optimal paging order is now determined by polling the cells in decreasing order of these residence probabilities [18].

So overall, the application of information theoretic methods to the location management problem allowed quantification of minimum network information flows to maintain accurate location information, provided an on-line method by which to characterize mobility, and in addition, endowed the paging process with an optimal paging sequence.

An extension of this information-theoretic framework for optimal location tracking in fourth generation (4G), heterogeneous (multi-system) wireless environment, encompassing multiple and possibly multi-tier wireless sub-networks (e.g., wireless LAN, CDMA network) has been recently proposed in [15]. While the optimal paging in a heterogeneous wireless network is an NP-hard problem [22], it is extremely challenging to determine the optimal trade-off between update and paging costs such that the information exchange (entropy) asymptotically approaches the rate distortion bound [21].

In the following subsection, we will consider other problems which can also benefit from the information theoretic framework.

3.2 Resource Management

Location uncertainty owing to node mobility leads to uncertainty in the availability of already scarce wireless resources such as bandwidth. We consider this type of resource uncertainty from the basic perspective of the channel unpredictability in cells, and perhaps most importantly from higher level perspectives of network resource management and associated quality of service (QoS) requirements. The resource provisioning at various layers of the protocol is thus a suitable mapping of joint uncertainty that propagates across layer boundaries.

3.2.1 Channel Prediction

The details of wireless link quality are often abstracted by Markov models of varying complexity based on the underlying fading channel models [16,28,29]. There have even been attempts to cobble such Markov models into protocols [12]. What is lacking is a general framework, especially for *ensembles* of channels which might be statistically related owing to such things as similar geography, similar mobility characteristics or mutual interference.

We can first ask whether there is some benefit to characterizing the entropy rates of channel processes to ascertain in some sense what a channel is "telling us" through a sequence of channel states. For instance, do the states of a given channel imply particular mobile trajectories, interference environments, weather/geographic conditions? What about relationships between the information provided by different channels? Are two channels mutually interfering? The answers to these sorts of questions can lead to better reactive control, but might also lead to development of network *services* which seek to coordinate use between co-located and mutual interfering nodes -- a sort of *spectrum server*.

We are currently examining through simulation and source data characterization (Lempel-Ziv, MaxEnt and others) whether channel state "itineraries" have structure which can be exploited for channel state prediction and ultimately better protocol design in much the same way that LeZi-update was developed and applied to node mobility.

This leads naturally to the notion of examining mutual information between different channel state sequences with an eye toward identifying nodes which could benefit from a spectrum coordination service.

3.2.2 Network Provisioning

A wireless system needs to be intelligent enough to make accurate predictions of node locations which will aid in pro-active reservation of wireless bandwidth, and development of efficient admission control mechanisms to guarantee QoS. It therefore seems obvious that reduction in node location uncertainty has the potential to reduce resource provisioning uncertainty.

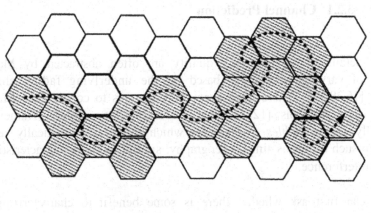

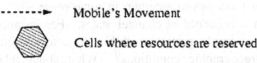

Mobile's Movement

Cells where resources are reserved

Figure 3 Resource Reservation along Node's Typical Route

Motivated by this idea, the concept of shadow cluster for a mobile terminal was introduced in [14] to identify a cluster of cells where the likelihood of finding the terminal is considerably high. Recently, mobility prediction based on LZ-compression has been used to dynamically form the shadow cluster around such a mobile terminal [26].

However, we suspect that better reservation schemes can be had by explicitly estimating the *most probable paths* a node is likely to follow in the near future. Specifically, of all the possible routes in a cellular network, a given node typically follows a small set of likely paths determined (for instance) by life-style or geography. Accurate knowledge of these paths and the ability to identify which path is currently being used by a given node could produce more accurate longer term predictions of resource demands therefore better resource scheduling. We are extending the concept of shadow clusters to one of *shadow routes* which can be learned using on-line algorithms such as LeZi-update and others. For details, refer to [8].

3.3 Traffic Load Characterization

The uncertainty of network traffic load has been well studied along with scaling phenomena that include self-similar characteristics [17]. Network resource provisioning with this type of traffic is nontrivial, particularly for wireless mobile networks. The concept of *effective bandwidth* [11] is traditionally used to map the traffic load uncertainty into network resource provisioning, where the effective bandwidth is defined as a statistical descriptor with *space* scale such as packet size and *time* scale such as packet arrival time. In an actual network, uncertainty in traffic load is captured by collecting a trace consisting of random sizes and arrival times of incoming packets.

The effective bandwidth based approach is found suitable for core networks in which a large number of *on-off* traffic sources are multiplexed together on a single stable link. However, in wireless access networks, each traffic source utilizes one unstable link and there are significant differences between the *on* (activity) and *off*(think) periods of individual sources. This can cause errors in effective bandwidth estimation. On the other hand, we find the concept of *useful data rate* [4], proposed for high-speed IP access networks, quite suitable for single on-off source and hence for wireless access networks.

Given the QoS requirements of a mobile node, we define the useful data rate as $a = V/T$, where V is the size of a file and T is the expected transfer time. Now the mobile's load in the wireless access network is nothing but a sequence of (empirically) measured useful data rates during an *on-off* period. To bring this empirical data into a symbolic domain, we assume that the measurement points are discrete with a constant separation in time scale. This assumption is justified by the fact that most of the current wireless standards use a fixed *time-slot* as the lowest unit to define the transport *frame* structure. With this definition, the traffic load of the mobile node at any instant can be represented by a string of symbol-pairs {0, a}, where "0" represents the null data rate per time-slot during the *off* period and "*a*" is the node's useful data-rate per time-slot during the *on* period. The requirement of *differentiated QoS* for different nodes will result in different strings of symbol-pairs such as {0, b}, {0, c}, etc. It is also possible to define multiple QoS classes for a given node.

For a given node, the string of symbol-pairs can be incrementally parsed and compressed using the LZ algorithm, based on the assumption that every node's activities and lifestyle exhibit some pattern of traffic load generation in the wireless network and this pattern can be learned on-line similar to the mobility tracking problem. The symbolic representation of traffic load permits the estimation of cumulative load in a cell with the help of joint and conditional probabilities of different nodes residing in that cell [1].

3.4   Network Topology and Routing

One of the most fundamental problems in the field of mobility management arises when network topology is variable owing to mobility or stochastic link variations so that routing tables are also stochastic. To efficiently route packets, nodes must have accurate information and the mobility management problem deals with how to obtain the information and distribute it where it is needed.

To date, this inherently information theoretic problem has not been considered analytically. Why? Perhaps because the issue is perceived as thorny owing to the variety of routing algorithms used for different purposes and the highly nonlinear mapping of node position and link quality to routing tables (for individual nodes and/or for global repositories). In fact, only recently have researchers begun to rigorously apply simple methods with an eye toward quantifying the amount of information necessary to specify network topology when nodes are mobile [23]. In many ways, consideration of this problem seems like a morass from information theoretic standpoint.

However, let us proceed undaunted and explore a bit further. To determine how much information a node might need to perform accurate routing, one can consider the entropy rate of the routing table as a lower bound on what a given node needs to be told in bits per second. But how does one map mobility to entropy rates of routing tables? To begin, we note that any transformation on information cannot increase information via the data processing theorem [7]. Viewed in that context, any deterministic transformation, nonlinear or not, cannot add additional information. So in the abstract, all that is necessary is a compact method of representing the various stochastic processes involved and specifying relationships.

Characterizing node position is reasonably simple and calculating entropy rates straightforward either analytically, numerically or through simulation. A routing table, however, seems to offer a particular difficulty since it is a collection of bits specifying instructions for packet delivery. Characterizing individual bit positions within tables seems unwieldy and fruitless, and relating routing tables of two nodes in a meaningful way even more so. However, information theory allows again a useful abstraction and, perhaps, a simpler way to consider the entropy rates of routing tables and even the information theoretic relationships between routing tables at different locations.

First, as an example, consider some distributed routing rule where nodes simply specify the next hop for incoming packets -- next hop routing. As before, under stochastic node mobility, each such routing table is a random process as well. We can then invoke the data processing theorem and note that any invertible function of a random variables does not reduce information. Thus, suppose we simply imagine a routing table at node i as a string of binary (± 1) values arranged in a vector v_i to which we apply an invertible, non-sparse, unitary matrix A, to obtain a vector $z_i = A v_i$. Since typically v has many components, each element of z_i will be approximately Gaussian, and Gaussians are easily and compactly characterized using covariance matrices and mean vectors. And given covariance, the entropy and entropy rates are easily derived. Furthermore, assuming mutually Gaussian random processes, we can now compare/relate the derived random variables $\{z_i\}$ of different nodes in a compact natural way as well via their covariances and thence by conditional entropies and entropy rates.

Now, certainly dimensionality could be a problem -- routing tables could be potentially very long strings of bits and the corresponding z_i vectors must have at least as many elements to avoid information loss. But once again, information theory provides a path for simplification. One would expect any individual routing table to be highly compressible, and any transformation which does not lose information can be applied without penalty. So one could imagine preprocessing routing tables (or even sequences of routing tables) using a standard Lempel-Ziv algorithm before applying the additive transformation which resulted in approximate Gaussians. One could also do a form of principal component analysis to derive a minimal set of uncorrelated Gaussian random variables from the jointly Gaussian ensemble. But the main point is that remarkably, even with such seemingly complex processing, the information theoretic relationships between the node mobility, the connection graph evolution and the routing tables have been maintained, rendered more tractable and perhaps most importantly from a validation/practical perspective, *measurable*. This last feature is especially important since it allows us build models of stochastic evolution of network connection graph and routing tables under mobility and/or link quality fluctuations. For details, see [19].

The information theoretic relationship between the processes at different nodes will also provide some measure of routing table redundancy between different nodes and aid in determining methods by which routing information can be efficiently disseminated over nodes -- and this is a first step toward answering the fundamental mobility management problem of who needs to know *what, where* and *when*. We expect these results to shed some light on the *how* as well.

4. CONCLUSION

We have proposed a fundamental overarching framework for wireless mobile networks to handle various uncertainties (channel, mobility, traffic load, resources, routing) in a *unified* manner. That is, once the uncertainty is couched in an information theoretic framework, one can develop methodologies and simulation tools which will strongly impact adaptive protocol design (e.g., resource scheduling) using the best possible information. We suspect this approach will in turn have tremendous impact on IP-convergent next generation wireless data network design and resource/capacity planning.

ACKNOWLEDGMENTS

We would like to thank Kalyan Basu and Amiya Bhattacharya for fruitful discussions. This work is partially supported by NSF ITR grant IIS-0326505.

REFERENCES

[1] K. Basu and S. K. Das, "Characterization of traffic load uncertainty in mobile networks," *Work in Progress*, 2004.

[2] A. Bhattacharya and S. K. Das, "LeZi-update : An information-theoretic approach to track mobile users in PCS networks," *Proc. ACM International Conference on Mobile Computing and Networking (MobiCom)*, pp. 1-12, Aug 1999.

[3] A. Bhattacharya and S. K. Das, "LeZi-update: An information-theoretic approach for personal mobility tracking in PCS networks," *ACM Wireless Networks*, vol. 8, no. 2, pp.121-137, 2002.

[4] T. Bonald, P. Olivier and J. Roberts, "Dimensioning high-speed IP access networks," *Proc. 18th International Teletraffic Congress*, Berlin, Germany, pp. 241-250, September 2003.

[5] A. Borodin and R. El-Yaniv, *Online Computation and Competitive Analysis*, Cambridge University Press, 1998.

[6] J. G. Cleary and I. H. Witten, "Data compression using adaptive coding and partial string matching," *IEEE Transactions on Communications*, vol. 32, no. 4, pp. 396-402, April 1984.

[7] T. Cover and J. Thomas, *Elements of Information Theory*, John Wiley, 1991.

[8] S. K. Das, A. Roy, K. Basu and A. Bhattacharya, "Resource reservation in wireless networks in the presence of uncertainty", *Work in Progress*, 2004.

[9] G. J. Foschini and M. J. Gans, "On limits of wireless communications in a fading environment using multiple antennas," *Wireless Personal Communications*, vol. 6, pp. 311-335, Mar 1998.

[10] E. T. Jaynes, "Papers on Probability, Statistics and Statistical Physics," Kluwer, 1989.

[11] F. P. Kelly, "Effective bandwidths at multi-class queues," *Queueing Systems*, vol. 9, no. 5, pp. 5-15, 1991.

[12] A. Konrad and B. Y. Zhao and A. D. Joseph and R. Ludwig, "A Markov-based channel model algorithm for wireless networks", *Wireless Networks*, vol. 9, pp. 189-199, 2003.

[13] S. Kullback and R. A. Leibler, "On information and sufficiency," *Ann. Math. Stat.*, vol. 22, pp. 79-86, 1951.

[14] D. Levine, I. Akyildiz and M. Naghsineh, "A resource estimation and call admission algorithm for wireless multimedia networks using the shadow cluster concept," *IEEE Transactions on Networking*, vol. 5, no. 1, pp. 1-12, Feb 1997.

[15] A. Misra, A. Roy, and S. K. Das, "An information-theoretic framework for optimal location tracking in multi-system 4G wireless networks" *Proc. IEEE INFOCOM*, Hong Kong, Mar 2004.

[16] H. S. Wang and N. Moayeri, "Finite-state Markov channel - a useful model for radio communication channels", *IEEE Transactions on Vehicular Technology,* vol. 44, pp. 163-171, Feb. 1995.

[17] H. Park and W. Willinger (Eds), *Self Similar Network Traffic and Performance Evaluation,* John Wiley, 2000.

[18] C. Rose and R. Yates, "Minimizing the average cost of paging under delay constraints," *Wireless Networks,* vol. 1, no. 2, pp. 211-219, July 1995.

[19] C. Rose, "An information theoretic framework for topology control and routing in wireless networks," *Work in Progress,* 2004.

[20] A. Roy, S. K. Das Bhaumik, K. Basu and S. K. Das, "Resource reservation for location-oriented multimedia in a smart home," *Proc. 8th Workshop on Mobile Multimedia Communications* (MoMuc), Munich, Germany, Oct 2003.

[21] A. Roy, A. Misra and S. K. Das, "A rate-distortion framework for information theoretic mobility management," *IEEE ICC,* June 2004.

[22] A. Roy, A. Misra and S. K. Das, "The minimum expected cost paging problem for multi-system wireless networks," *Proc. Workshop on Modeling and Optimization in Mobile, Adhoc and Wireless Networks* (WiOpt), pp. 94-103, Mar 2004.

[23] S. D. Servetto and G. Barrenechea, "Constrained random walks on random graphs: Routing algorithms for large scale wireless sensor networks," *Proc. ACM MobiCom,* 2002.

[24] C. E. Shannon, "A mathematical theory of communication," *Bell System Tech. Journal,* vol. 27, pp. 379-423, 623-659, 1948.

[25] C. Shannon, D. Moore and K. C. Claffy, "Beyond folklore: Observations on fragmented traffic," *IEEE/ACM Transactions on Networking,* vol. 10, no. 6, pp. 709-720, December 2002.

[26] F. Yu and V. Leung, "Mobility-based predictive call admission control and bandwidth reservation in wireless cellular networks," *Computer Networks,* vol. 38, pp. 577-589.

[27] J. Ziv and A. Lempel, "Compression of individual sequences via variable-rate coding," *IEEE Trans. Information Theory,* vol. 24, no. 5, pp. 530-536, 1978.

[28] M. Zorzi, R. R. Rao and L. B. Milstein, "On the accuracy of a first-order Markov model for data transmission on fading channels," *ICUPC* Nov. 1995.

[29] M. Zorzi and R. R. Rao and L. B. Milstein, "A Markov model for block errors on fading channels," *PIMRC'96,* Taiwan, Oct 1996.

Chapter 13

ROAMING IN THE GLOBAL WIRELESS INTERNET

CHIP ELLIOTT
BBN Technologies, 10 Moulton Street, Cambridge, MA 02138, USA

Abstract: Roaming across wireless networks and providers works reasonably well in the world's cellular systems but not in the Internet. That's not entirely surprising since the Internet was never designed to accommodate mobile endpoints. Despite a decade of hope for Mobile IP as the Internet's new mechanism for mobility, it has not yet gained popular acceptance. This paper proposes a more radical approach to Internet mobility, one that fundamentally severs the traditional connection between a host's IP address and its session endpoint identifier, thus allowing a computer to move from network to network, and provider to provider, without disrupting its active sessions. This approach can be viewed as a marriage of cellular mobility techniques with the Internet Protocol suite, and may leverage existing Internet mechanisms for access control and cellular mechanisms for roaming agreements. For clarity, we present a notional sketch of the resultant IPv6 mobility architecture.

1.   INTRODUCTION AND RELATED WORK

For many years it has been possible for cellular telephony subscribers to "roam" outside their home network, i.e., to use their cell phones outside their provider's network. To this end, a number of technical issues have been tackled and resolved (e.g. subscriber authentication, billing, etc.).

It is desirable for similar services to work with Internet devices, e.g., so that a visitor can plug a laptop into one's office network, so one can conveniently use local 802.11 services, and so forth. Indeed, it would be useful if mobile devices could "roam" from one access technology to another as needed, so that one could use an 802.11 LAN when in range but fall back to wide-area cellular services as needed.

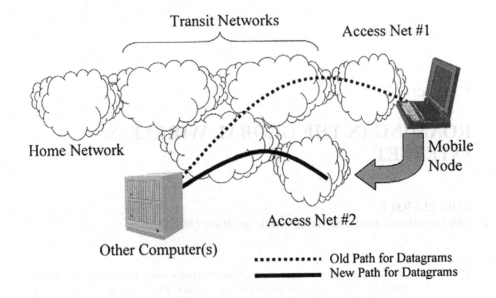

Figure 1 Internet roaming (Layer 3 handoff).

Fig. 1 shows such Internet roaming, and indeed of any form of "Layer 3 handoff," in highly schematic form. Here we see the most demanding form of roaming, in which a mobile node moves from one access network to another (perhaps run by different service providers) while maintaining an unbroken session – such as web browsing or IP telephony – to another computer. This has been difficult for the Internet Protocol suite, since a device's IP address incorporates its current subnet address. Thus when a device moves, its IP address changes which disrupts all its ongoing sessions.

It has long been expected that Mobile IP[1] should be the Internet's solution to roaming. However, Mobile IP adoption has been slow, perhaps because of the complexity of the tunneling and security interactions required to map between home and "care of" IP addresses. While Mobile IPv6 is somewhat simpler than Mobile IPv4 because IPv6 hosts are required to handle mobility (and thus issues with backwards compatibility have been solved by fiat), Mobile IPv6 is still a rather complex protocol suite.

In this paper, we argue that the key problem with IPv6 roaming lies in the semantics of IPv6 addresses themselves. A global, unicast IPv6 address[2,3] contains routing information and serves as the unique endpoint identifier for a transport session, e.g., IPsec or TCP. This arrangement is consistent with Internet history but works poorly with mobile host devices. Thus we take the bull by the horns, and suggest a redefinition of the IPv6 address space.

In particular, we note that the IPv6 address space is big enough to segment a 128-bit address into a globally unique endpoint identifier (64 bits) prefixed by one or more routing headers (64 bits apiece), which segmentation makes roaming much easier. Although independently conceived, this approach has similarities with so-called "8+8" schemes which also segment an IPv6 address into 64-bit routing information and 64-bit unique identifier. See O'Dell[4] and others[5] for interesting earlier work on 8+8 addressing architectures. Both this earlier work and ours have been influenced by the Nimrod[6] routing and addressing architecture.

Needless to say, cellular telephony systems have also been a fruitful source of techniques. Our treatment of mobile identification numbers and roaming protocols has been influenced by GSM approaches[7]. Indeed, one can view this paper as an attempt to port proven ideas from cellular telephony into the Internet protocol suite.

2. KEY IDEAS IN THIS APPROACH

Key ideas in this approach fall into two major categories, those which are shared with other 8+8 approaches:

- Split the IPv6 address into an 8-byte routing prefix, followed by an 8-bye globally unique endpoint identifier
- Base session endpoints on the 8-byte unique endpoint identifier, rather than the full 16-byte IPv6 address

And those which appear to be new:

- A simple strawman architecture for Internet mobility
- Discussions of AAA and/or IPv6 address support for Internet roaming
- Subsequent sections should make these key concepts clear in practice.

3. SYSTEM DESIGN AND MAJOR ELEMENTS

Fig. 2 displays the basic system elements involved in this approach to Internet roaming. The Mobile Node (M) is an Internet host using the IPv6 techniques described in this paper. Its own Home Network is shown to the left, along with at least one Domain Name Server (DNS) and Registry operated by that network's provider.

AAA #1 and #2 are Authentication, Authorization, and Accounting servers such as are typically used for controlling Internet access[8]. AAA #1 may be run by one service provide, #2 by another.

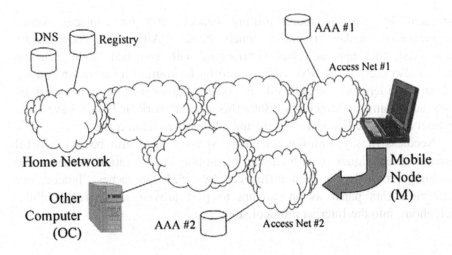

Figure 2 Basic system elements in Internet roaming.

Other Computer (OC) stands for any computer in the Internet which may wish to communicate with M (and thus which needs to find M when M is roaming), or which has an active session with M. The various networks shown may be public or private, e.g., they may be operated by Internet Service Providers (ISPs), corporations, home users, etc.

4.   STRAWMAN IPV6 ADDRESS PLAN

When an Internet host roams into a new access network, it must supply enough information to the local AAA server so that server can find the host's home registry. This information can be included directly in the host's IPv6 address, or alternatively it may be supplied at a higher layer in the protocol stack. We consider each approach in turn.

Fig. 3 presents a possible new IPv6 address plan for mobile devices, or more specifically, for a network interface on such a device. It consists of four major segments: an address type field (FP), reserved bits (RES), a home registry ID, and a unique ID for that interface. The FP field's value (010) is different from any IPv6 address prefix currently defined. This value indicates that the remainder of the IPv6 address is formatted as per discussions below. Such address formats would only be used when a device (interface) is first attempting to gain access to a network, i.e., it is used only during the bootstrap phase of gaining network access.

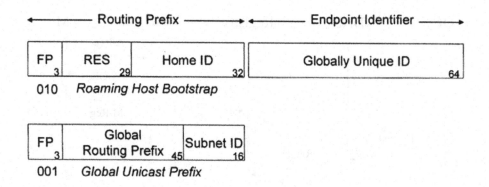

Figure 3 Possible Address Formats for IPv6 Mobile Device.

Reserved bits (RES) are for future use, and must be zero. The Home ID field provides a concise shorthand for finding the device's home registry. The Unique ID is a 64-bit long string that uniquely identifies this network interface device; for example, it might be an IEEE-defined EUI-64 identifier[9] for a particular 802.11 device.

When a network AAA server is given such an address, it has all the information it needs to find the Home Registry for this device, and hence to perform any accounting protocols that may be required before access can be granted. This process is conceptually very similar to that used in cellular telephony and so is not described here. (Please see a good tutorial[4] for details.) Once it has completed its transaction with this home registry, it can then supply a Global Unicast routing prefix to the mobile host, which it may then use while connected to this local access network.

Alternatively, the roaming host may use an IPv6 address valid only for local communications, and then convey its registry information via a higher layer of the protocol stack, e.g., as a DNS name conveyed within the AAA protocol between the roaming host and the local AAA server.

5.   ROAMING AND REGISTRATION PROTOCOLS

Fig. 4 depicts sample host protocol interactions that occur when a mobile device (M) roams into a new network, and then is contacted by some other computer (OC).

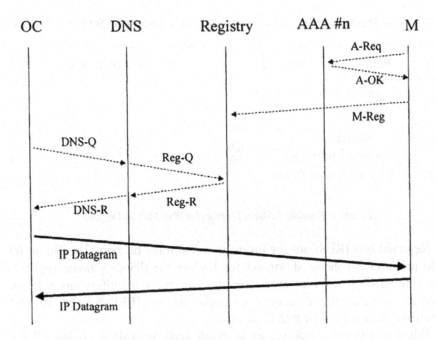

Figure 4 Example protocol interactions during roaming.

First, M performs typical access request protocols with the AAA server for its new network. M uses its globally unique IPv6 Mobile Device Address (Fig. 2) for these interactions. These interactions are drawn as A-Req and A-OK in Fig. 4, though of course greater protocol complexity is possible. When AAA #n grants entry to M, it provides one or more global routing prefixes for M in its A-OK message, and opens its firewalls so that M can then communicate globally using these prefixes.

Second, M formulates a globally routable unicast address using one of these prefixes, and sends this latest address information to its home registry. This interaction is shown as M-Reg in Fig. 4; in reality it is likely to involve IPsec, some form of authorization, and so forth, but purely between M and its home Registry.

Later, OC attempts to find M for communication. It performs a DNS query (DNS-Q) and obtains M's current set of globally routable unicast addresses in the response (DNS-R). Here we have shown it using M's home DNS, and that DNS performing a look-aside operation to the local Registry to find M's current addresses (that is, its current routing prefixes and its unchanging unique identifier). The DNS and Registry servers could be combined if desired, and more complex caching schemes, perhaps with invalidation upon roaming, could also be feasible. We show the simplest possible scheme for clarity; clearly this mechanism is just a form of Dynamic DNS.

Once OC has retrieved one or more IPv6 addresses for M from its DNS lookup, it communicates directly with M.

The provider-to-provider interactions are not shown on Fig. 4 but are straightforward, as mentioned at the end of the previous section. Once AAA #n has received A-Req from M, it determines M's home registry, and performs accounting protocols to ensure that M is still a customer in good standing, roaming agreements are in place, etc. If all is well, it then responds with A-OK; otherwise it may refuse M's network access. Of course non-profit access networks may skip some or all of these accounting steps.

Finally we note that providers may choose to work through third-party clearing houses for their roaming transactions. It should be evident that this too can be easily obtained by this approach, which is, after all, rather similar to those already well understood for cellular telephony.

6.   THE COSTS OF ROAMING

There are four basic costs for roaming with this proposal: network access, accounting transactions, registry updates, and DNS lookups. We consider each in turn.

Network access costs count the number of AAA transactions. This depends on the number of roamers into a given network but in general is likely to be fairly low compared to the number of native devices, and to consume about the same resources as AAA for a native device. Thus if an access network has as many active roaming devices as active native devices, it will at most need to double its number of AAA servers.

Accounting transactions scale with the number of network access attempts. Let us consider a metropolitan wireless network with 10 million subscribers and 10 million roaming devices. If all roaming devices connect one morning (e.g. during 5 hours), this averages 2 million transactions / hour, or 555 / second, which is not a particularly demanding load with today's database technology, and could probably be accommodated by one or two servers. In the same scenario, a billion roaming devices would require perhaps a hundred accounting servers. If desired, caching could further reduce this load.

Registry updates scale similarly to accounting transactions, and so they too are likely to require no more than a few hundred servers for even billions of roaming devices.

Finally, DNS updates scale with the number of requests. In the simplest approach, device mobility defeats DNS caching since one always needs the latest IPv6 address for a device; a cached address may be outdated. It is hard to estimate the number of lookups / second that will be performed in order to contact a mobile device (at present, most servers are at fixed locations but IP telephony has different usage patterns), but roaming cellular telephones may present a useful example. If one assumes 1 billion devices, of which 10% are actively roaming at any time (100 million), and each one is contacted every 10 minutes, then 166,666 lookups are performed every second. If a server can support about 10,000 requests / second, this requires 17 servers.

All told, then, every very large populations of roaming devices could be supported by at most hundreds of servers.

7. SESSIONS AND ENDPOINT IDENTIFIERS

It should be evident that this scheme allows a mobile device M to move between access networks freely, with a separate global unicast routing prefix for each network in which it is currently participating. Indeed M may participate in a number of access networks at the same time, e.g., an 802.11 LAN and a 3G cellular system.

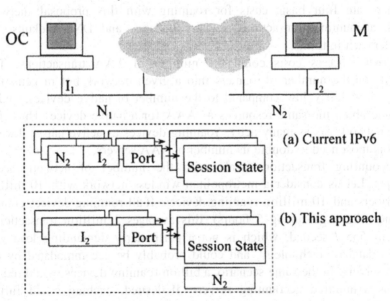

Figure 5 Session identifiers based on globally unique identifiers.

Fig. 5 shows how session state may be stored in OC so that an active session with M may be maintained across network handoffs, roaming, etc. As previously mentioned, this is accomplished by indexing session state to the globally unique 64-bit identifier for M's network interface device rather than its full 128-bit IPv6 address.

As shown, the current definition of IPv6 indexes the session state by M's full IPv6 address, including a subnet ID (N2), an interface number (I2) which identifies a given network interface on the network, and a port number. Because IPv6 does not require interface numbers to be globally unique, the full IPv6 address must be used in order to derive a unique identifier for a session endpoint.

Contrast that with the approach outlined in this paper, where I2 itself is a globally unique identifier, again tagged with the appropriate port number. A set of routing prefixes N2 etc. may then be stored in the session database, so that any of them may be used in order to transmit datagrams in the session. Note that session datagrams may be received from any routing prefix. As in conventional Mobile IP, control fields may be employed in datagrams (or separate control messages) so that M can inform OC as new routing prefixes are acquired and old ones become obsolete.

Alternatively, session endpoints could be based upon a given node's own identity rather than that of one of its interface devices; for example, I2 could be a unique CPU serial number. Such a scheme would allow seamless handoff across differing devices, e.g., from 802.11 to cellular systems.

This approach to session identifiers could be applied across the Internet Protocol suite, e.g., for use with TCP, RTP, and of course IPsec. (Its extension to IP multicast is left as an exercise for the reader.)

8.   SUMMARY

We argue that Internet roaming and other forms of "Layer 3 handoff can be managed in a fairly simple fashion, if one is willing to change the Internet architecture in two ways: (a) split the IPv6 address into an 8-byte routing prefix and a globally unique 8-byte interface identifier; and (b) base all session endpoints on the 8-byte unique endpoint identifier rather than the full IPv6 address. A strawman architecture shows that Internet roaming and mobility are relatively easy to manage in this scheme, with need for any complexities such as tunnels or encapsulation.

AKNOWLEDGEMENTS

The author offers his sincere thanks to Dr. Raj Ganesh for his kind words and for his patience as this paper slowly came into existence.

REFERENCES

1. H. Soliman, Mobile IPv6 : Mobility in a Wireless Internet, Addison-Wesley, Reading MA, 2004.
2. R. Hinden, S. Deering, "Internet Protocol Version 6 (IPv6) Addressing Architecture", RFC 3513, April 2003.
3. R. Hinden, S. Deering, E. Nordmark, "IPv6 Global Unicast Address Format", RFC 3587, August 2003.
4. M. O'Dell, "8+8 – An Alternate Addressing Architecture for IPv6", expired Internet Draft, October 1996.
5. M. Sola, M. Ohta, Y. Muraoka, T. Maeno, "The 8+8 IPv6 Addressing Architecture", poster in INET 2000, Yokohama, Japan, Internet Society CD Proceedings, June 2000.
6. N. Chiappa, "IPng Technical Requirements of the Nimrod Routing and Addressing Architecture," RFC 2260, December 1994.
7. A. Mehrotra, GSM System Engineering, Artech House Publishers, Boston, 1997.
8. See for example the IETF's web site for AAA, http://www.ietf.org/html.charters/aaa-charter.html.
9. IEEE, "Guidelines for 64-bit Global Identifier (EUI-64) Registration Authority", http://standards.ieee.org/regauth/oui/tutorials/EUI64.html, March 1997.

Chapter 14

PROXY SERVICES FOR THE MOBILE INTERNET

VICTOR C.M. LEUNG

The University of British Columbia, Vancouver, BC, Canada V6T 1Z4

Abstract: The mobile Internet employs various wireless networks to enable mobile clients to access Internet services. As wireless links generally suffer from limited bandwidth and higher error rates when compared to wired links, the performance of Internet access is often less than ideal. A common approach to overcome this problem is to use a proxy to enhance the performance of the mobile Internet. This paper provides an overview of a number of proxy services being developed at the University of British Columbia under the "Multimedia content delivery over the mobile Internet" project. These services include header/data compression for wireless application protocol, vertical handover support between cellular and wireless local area networks, and multi-channel transport layer security.

1. INTRODUCTION

Recent advances of wireless data networking technologies such as 2.5 and 3G cellular networks as well as 802.11 wireless local area networks (WLANs)[1] make it attractive to extend Internet services to people on the move, who can access the Internet via mobile clients equipped with appropriate wireless network interfaces. Although bandwidths provided by wireless data networks have seen substantial improvements in the last few years, compared to wireline facilities wireless links still suffer from lower bandwidths and higher error rates. For example, it is well known that these limitations result in performance degradations of the end-to-end transmission control protocol (TCP), which misinterprets loss of packets due to errors over a wireless link as congestion and unnecessarily reduces the transmit window, thus lowering the throughput of the connection[2]. A suitable solution to this problem is to use a performance enhancement proxy[3] to break the end-to-end TCP connection into two indirect-TCP connections over the wireless and wired segments, so that the limitations of the wireless link can be properly addressed over the wireless domain. This approach is also reflected in the architectural design of the wireless application protocol (WAP)[4], which employs a proxy gateway as shown in Figure 1 to convert text pages written in wireless markup language (wml) to bandwidth efficient byte codes in version 1.x. Although WAP version 2[5] supports wireless-profiled TCP[6], a proxy gateway is still needed to facilitate interworking with terrestrial TCP connections that do not support the wireless profile. Until such time that the protocols in the Internet have evolved to fully embrace wireless networks and mobility, and these new protocols have been widely adopted, there is an ongoing need for proxy services to harmonize the differences in quality of service offered by wired and wireless networks.

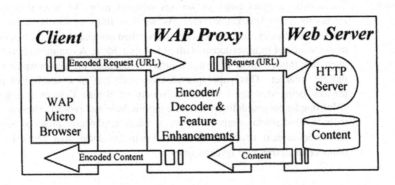

Figure 1 WAP proxy model

Sponsored by Telus Mobility, the Advanced Institute of BC, and the Canadian Natural Sciences and Engineering Research Council, our research group at UBC has been developing various proxy services for the mobile Internet over the past few years under the "Multimedia content delivery over the mobile Internet" project. This paper gives an overview of some results of completed and ongoing research. In section II, we present a proxy service for WAP 2.0 that provides advanced header/data compression at the TCP and IP layers. In section III, a proxy service that employs the Stream Control Transfer Protocol (SCTP) to enable seamless vertical handoffs between cellular networks and WLANs is described. Section IV presented the multi-channel secure socket layer (MC-SSL) that extends configurable end-to-end security across proxies. Section IV concludes the paper and describes future research directions.

2. DATA COMPRESSION PROXY FOR WAP 2.0

2.1 Motivations and objectives

Developed by the WAP Forum, WAP[4] is a de-facto world standard for the delivery and presentation of wireless information services to mobile phones and other wireless terminals via cellular wireless networks. Compared to WAP 1.x, which employs wml-coded contents transferred by byte codes over the wireless links, WAP 2.0[5] is a positive step forward towards convergence of the mobile Internet with the global Internet as it supports XHTML and wireless-profiled HTTP and TCP. However, as discussed above, the use of a proxy gateway is still needed to facilitate access to many servers that may not support wireless profiled HTTP and TCP. On the other hand, increased compatibility with the Internet is obtained at the cost of losing communications efficiency over the wireless links, due to the large amount of overhead in IP packets. The objective of this research project is to compare the data transfer efficiency of WAP 2.0 with WAP 1.x over emulated IS-95 and cdma2000 1xRTT data links, and to develop a data compression proxy architecture to enhance the efficiency.

2.2 Proposed architecture

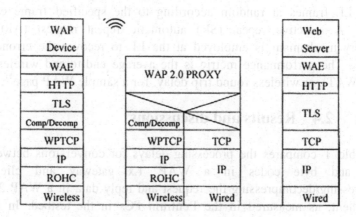

Figure 2 Data compression architecture for WAP 2.0

We propose a data compression proxy architecture for WAP $2.0^{7,8}$ as shown in Figure 2, which applies Robust Header Compression (ROHC)[9] to TCP/IP headers, and lossless compression to TCP payload. This approach allows preservation of end-to-end data security through Transport Layer Security (TLS) tunneling, independent of whether or not conversion between the standard version of TCP and wireless-profiled TCP is performed at the proxy. We consider compression of the HTTP replies only, as well as compression of both HTTP requests and replies.

2.3 Performance evaluations

System performance was evaluated using a testbed, which consists of a WAP server, a WAP proxy (WAP gateway for WAP 1.x), a client, and an emulated wireless channel between the proxy and client, implemented in four PCs connected via a private Ethernet. The emulation function of the ns-2 simulator[10] is used to emulate the packet level behaviours of IS-95 and cdma2000 1XRTT wireless channels. Emulation refers to the ability to incorporate real-time simulated behaviour into a live network. The emulated channel consists of two nodes: a mobile node connected to the mobile client, and a base station node connected to the proxy. The transmitted IP packets between the mobile client and the proxy are captured by the simulator, and then fragmented into Link Layer (LL) frames according to the system parameters. The fragmented LL frames are sent every 20ms to the other node. When all fragments of an IP packet are received at the other end, the IP packet is reassembled and injected back to the live traffic. The maximum user data transmission rate for IS-95 is 9.6 Kbps, and 153 Kbps for CDMA2000 1XRTT. Channel error is modelled by a function at LL that drops LL frames at random according to the specified frame error rate (FER). A selective repeat (SR) automatic repeat request (ARQ) error recovery mechanism is employed at the LL to recover the erroneous LL frames. The performance metric is the average end-to-end wireless access time (WAT), or wireless round trip delay, for a sample WAP page[11].

2.4 Results and discussions

Table 1 compares the processing delays for conversions between wml texts and byte codes in a WAP 1.x gateway and client, and compressing/decompressing the request and reply data in a WAP 2.0 proxy and client, as measured in the Pentium PCs in the testbed. In practical implementation a shorter delay at the proxy/gateway is expected, while a longer delay is expected at the mobile client. These delays and the actual transmission times are key factors in the different WAT shown below.

Table 1 Processing delay in WAP proxy and client

WAP 1.x	WAP 2.0 (HTTP, TCP/IP)		
	No compress	Reply compress	Request & Reply compress
180.05ms	4.02ms	6.69ms	11.51ms

In tests over an emulated IS-95 channel with a data rate of 9.6 Kbps, results in Figure 3(a) show that the WAT of WAP 1.x is 20-25% of that of WAP 2.0 employing TCP/IP without data compression, but applying request and reply compression and ROHC to WAP 2.0 reduces the WAT to approximately the same level as WAP 1.x. This shows the advantage of using WAP 1.x over low-bandwidth networks and the needs for data and header compression when using WAP 2.0 over such networks. The performance gains achieved by different elements in the compression scheme are shown in Figure 3(b). Compared with the WAT for WAP 2.0 without compression shown in Figure 3(a), Figure 3(b) shows that compressing reply data only reduces WAT by 66.5% at 1% FER to 76.2% at 40% FER. Combining request and reply compression achieves a 4% reduction over reply compression. However, WAP 1.x still performs better than WAP 2.0 with content compression only. Adding ROHC gives a 12% reduction in WAT over content compression, yielding performance slightly better than WAP 1.x. Overall, WAP 2.0 with request/reply compression in combination with ROHC reduces WAT by 71.6% at 1% FER and 79.8% at 40% FER over WAP 2.0 with no compression.

Over an emulated CDMA2000 1XRTT channel with a data rate of 153 Kbps, WAP 2.0 performs better than WAP 1.x even if no compression is applied, with 32.5% and 9% lower WAT at 1% and 40% FER, respectively, as shown in Figure 4. This is because of the large protocol conversion delay in the WAP 1.x gateway and client compared to the transmission time. Applying reply compression with ROHC further reduces the WAT of WAP 2.0 by more than 50% compared to no compression; most of this attributed to content compression in the reply packets. Interestingly, due to the short packet lengths, compression of request packets increases the overall WAT compared to reply compression as the reduction in transmission time is more than offset by the increase in processing time. When FER is high, increased packet retransmissions in WAP 2.0 without compression result in steep increase in WAT; however, content compression helps to reduce the number of packets transmitted and partially offsets the effect of high FER.

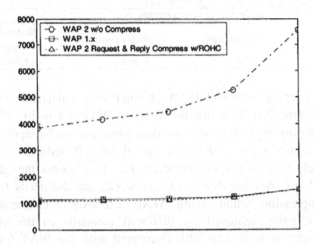

(a) WAP 2 without compression compared with others

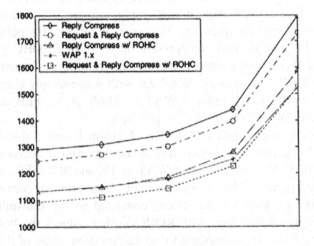

(b) Contributions of different compression schemes

Figure 3 WAP Performance over IS-95 channel

The results show that WAP 1.0 is more suitable for low-speed wireless networks, while WAP 2.0 is more suitable for the high-speed wireless networks. In all cases, using the proposed compression scheme that combines robust header compression with data content compression can substantially reduce the wireless access time.

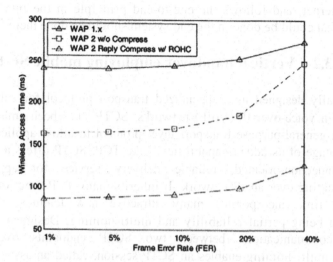

Figure 4 WAP Performance over CDMA2000 1XRTT channel

3. SCTP FOR SEAMLESS VERTICAL HANDOVER

3.1 Motivations and objectives

WLANs and 2.5/3G cellular networks are complementary in that the former offer a high data rate but only tolerate low mobility, whereas the latter supports high mobility at a reduced data rate. Therefore integration of and interworking between WLANs and cellular networks are topics of significant current research interest[12], and in fact seamless integration of these networks is considered a key characteristics of the next generation wireless networks[13]. While mobile-IP[14] provides IP-layer support of node mobility across heterogeneous networks and is the preferred technique for interworking cellular and WLANs[15,16], it requires additional mobility-support infrastructure such as home agents and foreign agents, and coordination between providers of the different wireless services. Our objective is to develop a technique for vertical handover (VHO) between cellular and WLANs that is seamless, i.e., with minimal or no disruption of ongoing communication sessions, and which requires no coordination between network operators and no additional mobility supporting network infrastructure other than a proxy server external to the wireless networks. The proposed method employs SCTP[17], a next generation transport protocol for the Internet, and follows the end-to-end principle in the Internet, i.e., anything that could be done in the end system should be done there[18].

3.2   Vertical handover employing mobile SCTP

Originally designed as a specialized transport protocol for call control signaling in voice-over-IP (VoIP) networks, SCTP[17] has been embraced by IETF as a general-purpose transport layer protocol to allow applications to take advantage of its added capabilities. Like TCP, SCTP offers a point-to-point, connection-oriented, reliable delivery service for applications communicating over an IP network. It inherits many TCP functions and at the same time incorporates many attractive new features, the most interesting being partial reliability and multi-homing. Designed for fault resilient communications between two SCTP endpoints over wired networks, multi-homing enables an SCTP session, called an association, to be established over multiple interfaces by specifying multiple IP addresses in the association. An SCTP association between two hosts, say, A and B, is defined as: {[a set of IP addresses at A] + [Port-A]} 4+ {[a set of IP addresses at B] + [Port-B]}. SCTP normally sends packets to a destination IP address designated as the primary address, but can reroute packets to an alternative, secondary IP address if the primary IP address becomes unreachable. This powerful feature has been exploited to support host mobility using SCTP. Specifically, the SCTP dynamic address reconfiguration (DAR) extension[19], called mobile SCTP (mSCTP)[20,21], can provide a simple but powerful framework for mobility support over IP networks. Here, we apply mSCTP to support VHO of a mobile client (MC) between a loosely coupled[22] WLAN and a Universal Mobile Telecommunications System (UMTS) as shown in Figure 5[23], where both the mobile client and the fixed server (FS) supports mSCTP. In practice the FS is likely to support only TCP, in which case FS can be replaced by a proxy server that provides the necessary SCTP terminations. Here we consider FS synonymous to a proxy server.

Using the multi-homing feature of SCTP with DAR extension, an MC can have an initial IP address from UMTS, an then dynamically add a new IP address obtained from a WLAN when coverage is available, then switch to the latter as the primary address. Similarly, when the MC leaves the WLAN coverage, it reverts back to the UMTS IP address as its primary address.

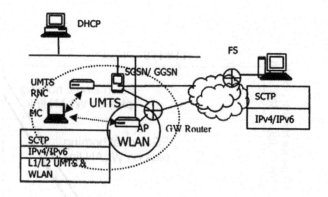

Figure 5 UMTS/WLAN VHO using mSCTP

3.3 Single versus dual-homing proxy servers

While the MC needs to have both WLAN and UMTS interfaces, which allow it to operate in dual-homing configuration, the FS can be either single- or dual-homing. We have developed detailed VHO procedures for both FS configurations[23,24]. By applying message bundling in the VHO signaling, a more efficient VHO process is realized for dual-homing FSs.

We evaluate the delay and throughput of UMTS to WLAN and WLAN to UMTS VHOs, using the ns2 simulator[10] and extending the SCTP module developed by the University of Delaware[25] to support multi-homing with DAR over wireless links. The bandwidths are set to be 384 kbps for the UMTS link and 2 Mbps for the WLAN link. The network propagation delay is set to 100 ms. FTP traffic is triggered from MC at time 1 s. VHO Triggering process is activated at time 5 s. Results shown in Figures 6 and 7 indicate that when dual-homing is available at the FS, the VHO delay is significantly reduced and a higher throughput can be maintained. However, as most servers are connected to the Internet using a single network interface, the use of a proxy server with dual network interfaces to support VHO would allow the advantage of a dual-homing FS to be readily realized.

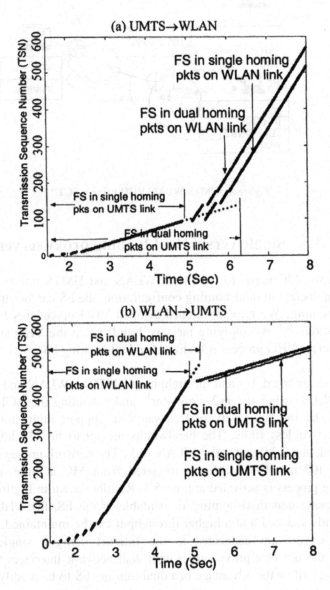

Figure 6 Delay Performance of vertical handover

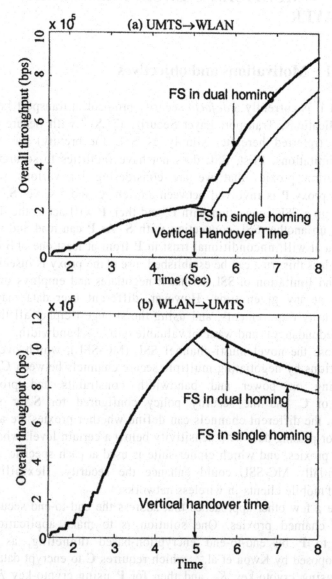

Figure 7 Throughput Performance of vertical handover

4. MULTICHANNEL SECURE SOCKET LAYER

4.1 Motivations and objectives

Although it is currently a *de facto* security protocol at transport layer for Internet applications, Transport Layer Security (TLS)[26] with Secure Socket Layer (SSL) (referred hereafter simply as SSL for brevity) has several functional limitations. First, SSL does not have facilities to securely deal with application proxies that we are considering for various wireless services: if a proxy **P** is involved between a client **C** and a server **S**, **C** will normally set up a SSL connection with **P**, and then **P** will act as the delegate of **C** and set up another SSL connection with **S**. As **P** can read and modify sensitive data at will, unconditional trust in **P** from at least one of **S** or **C** is required. Unless this trust can be established, use of the proxy is insecure.

The second limitation of SSL is that it negotiates and employs only one cipher suite at any given time. However, different user data may have different security requirements, and using the strongest cipher all the time can lead to redundancy[27] and waste of valuable wireless bandwidth.

We propose the novel multi-channel SSL (MC-SSL), which overcomes these limitations by negotiating multiple secure channels between **C**, **S** and **P**. Depending on power and bandwidth constraints and processing capabilities of **C** and the security policy configured for **S** or specific applications, the different channels can define whether proxies are allowed to be used for passing data with sensitivity below a certain level, what types of data need proxies, and which cipher suite is used at each specific security level. Potentially MC-SSL could enhance the security, flexibility and efficiency of mobile clients in wireless networks.

There are a few other approaches that address the end-to-end security for the case of chained proxies. One solution is to make application data unreadable to **P** by end-to-end encryption-based tunneling, as in the approach proposed by Kwon et al.[28], which requires **C** to encrypt data twice: first for **S** using crypto-key K_S, and then for **P** using crypto-key K_P. The drawback of this method is that application layer functions such as content transformation and virus scanning cannot be performed by **P**.

4.2 High level description of MC-SSL

MC-SSL uses a multiple-channel model, in which each channel can possess its own characteristics including cipher suite and data flow direction.

A special type of channel called the proxy channel enables MC-SSL to support partially trusted proxies. In the conceptual proxy model of MC-SSL shown in Figure 8, **C-S** is an end-to-end channel, and **C-P-S** is a proxy channel that relies on the **C-S** channel to control channel negotiation and application data delivery. Thus, **C** and **S** can deliver data over **C-S** or **C-P-S**, according to the requirements for proxy and the sensitivity of data. As a result, sensitive data, such as id/password or credit card number, do not have to be exposed to **P**. A MC-SSL session may negotiate zero or more proxy channels. Each proxy channel and the corresponding end-to-end channel form a triangular relationship with the proxy as the third vertex.

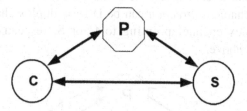

Figure 8   Triangular proxy model of MC-SSL

In SSL, a cipher suite consists of a key exchange algorithm, a cipher, and a hash algorithm used to compute Message Authentication Code (MAC). In MC-SSL, a cipher suite consists of only two elements: {cipher and key size, hash algorithm for MAC}. A MC-SSL connection can have multiple cipher suites. A point-to-point connection can be characterized as: {point 1, point 2, key exchange algorithm, {cipher suite 1, cipher suite 2, ...}}, where each cipher suite forms a channel. Every MC-SSL connection must first negotiate a cipher suite strong enough to form the primary (or backbone) channel, which is employed to set up and control other (secondary) channels. Figure 9 illustrates a connection between A and B characterized by {A, B, RSA, {CS1, CS2, CS3, CS4}}, where RSA is the key exchange algorithm, and CS1 to CS4 are four different cipher suites. The primary channel, channel 1, employs CS1.

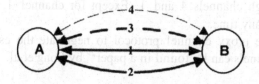

Figure 9   Multiple cipher suites in point-to-point connection

A combination of the proxy model and the multiple cipher suites yields the multiple-channel model illustrated in Figure 10. In MC-SSL, a channel can be defined as a virtual communication "pipe" with or without intermediate application proxies. Two MC-SSL endpoints communicate with each other through the pipe using a cipher suite. In addition, a channel can be either duplex, or simplex with a flow direction. A MC-SSL channel is thus characterized by the following attributes: {*channel id, endpoint1, endpoint2, proxy, direction, cipher suite*}, where *channel id* is the identifier of a channel in a MC-SSL session context, *endpoint1* and *endpoint2* are either DNS names or IP addresses of corresponding machines, *proxy* is null if the channel is end-to-end, or else it is the DNS name or IP address of the proxy in a proxy channel, *direction* can be D for a duplex channel, or C or S indicating a simplex channel pointing to **C** or **S**, respectively, and *cipher suite* is as defined above.

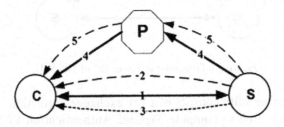

Figure 10   Multiple-channel model of MC-SSL

In Figure 10, the MC-SSL session has five channels, where channels 1 and 4 are primary (or backbone) channels, and channels 2, 3, and 5 are secondary channels. In addition, only channel 1 is a duplex channel for application data; others are simplex channels from **S** to **C**. Typically, **C** can use channel 1 to send encrypted requests to **S**, and **S** can choose one of the five channels to send back the responses according to the contents. These channels are established in the following order: channel 1 is negotiated first; channels 2 and 3 are negotiated through channel 1; channel 4 is the first proxy channel and is also negotiated through channel 1; channel 5 is negotiated through channels 4 and 1. Except for channel 1, other channels can be set up at any time.

Design of the proxy channel protocol to negotiate the establishment of the primary channels can be found in a paper[29] by Song et al.

5. CONCLUSIONS AND FUTURE WORK

We have presented the design and analysis of a number of proxy services for the mobile Internet, including header/content compression, vertical handover support, and multi-channel security, that together enable more efficient, flexible and secure data transfer over heterogeneous wireless networks. Some of these proxy services are currently being implemented for proof-of-concept. We are also extending our research on proxy services for the mobile Internet in several different directions. First, we are developing content transformation and scalable content creation techniques for world wide web and multimedia contents that can present contents based on the capability of the mobile client and the wireless channel. Second, we are developing techniques to enhance the performance of SCTP over wireless networks and to better adapt to changes in network bandwidth during vertical handovers. Third, we are addressing issues of interworking SIP and MC-SSL with SCTP to supporting multimedia connections over the mobile Internet.

AKNOWLEDGEMENTS

The work presented in this paper has resulted from collaborations with several colleagues and graduate students, including Drs. Konstantin Beznosov, Tejinder Randhawa and Fei Yu, and Li Ma, Yong Song and Zhanping Yin. We gratefully acknowledge the support of this work by Telus Mobility, the Advanced Systems Institute of BC, and the Natural Sciences and Engineering Research Council of Canada through grant CRD247855-01.

REFERENCES

1. J. de Vriendt, P. Lainé, C. Lerouge and X. Xu, Mobile network evolution: a revolution on the move, *IEEE Commun. Mag.* **40**(4), 104–111 (2002).
2. H. Elaarag, Improving TCP performance over mobile networks, *ACM Computing Surveys* **34**(3), 357–374 (2002).
3. J. Border, M. Kojo, J. Griner, G. Montenegro and Z. Shelby, Performance enhancing proxies intended to mitigate link-related degradations, *IETF RFC 3135,* Jun. 2001.
4. WAP Forum, WAP architecture specification, version 12-July-2001, http://www.wapforum.org/what/technical.htm.
5. WAP Forum, WAP 2.0 technical white paper, Jan. 2002.
6. WAP Forum, Wireless profiled TCP, version 31-March-2001, http://www.wapforum.org/what/technical.htm.
7. Z. Yin and V.C.M. Leung, A proxy architecture to enhance the performance of WAP 2.0 by data compression, in *Proc. IEEE WCNC,* New Orleans, LA, Mar. 2003.

8.   Z. Yin, *A proxy architecture to enhance the performance of WAP 2.0 by data compression,* MASc thesis, University of British Columbia, 2003.
9.   C. Bormann *et al.,* Robust header compression (ROHC), *IETF RFC3095,* July 2001.
10.   http://www.isi.edu/nsnam/ns.
11.   B. Eged, T. Dezso and F. Egedi, Server side round-trip delay measurements in WAP environments, in *Proc. IEEE IMTC,* Budapest, Hungary, pp. 525–529, May 2001.
12.   V.K. Varma, K.D. Wong, K.-C. Chua and F. Paint, Guest editorial - Integration of 3G wireless and wireless LANs, *IEEE Commun. Mag.* **41**(11), 72–73 (2003).
13.   S.Y. Hui and K.H. Yeung, Challenges in the migration to 4G mobile systems, *IEEE Commun. Mag.* **41**(12), 54–59 (2003).
14.   C. Perkins, IP mobility support, *IETF RFC2002,* Oct. 1996.
15.   M.M. Buddhikot, G. Chandranmenon, S. Han, Y.-W. Lee, S. Miller and L. Salgarelli, Design and implementation of a WLAN/CDMA2000 interworking architecture, *IEEE Commun. Mag.* **41**(11), 90–100 (2003).
16.   J.W. Floroiu, R. Ruppelt, D. Sisalem and J.V. Stephanopoli, Seamless handover in terrestrial radio access networks: A case study, *IEEE Commun. Mag.* **41**(11), 110–114 (2003).
17.   R. Stewart, *et al.,* Stream Control Transport Protocol, *IETF RFC2960,* Oct. 2000.
18.   J.H. Saltzer, D.P. Reed and D.D. Clark, End-to-end arguments in system design, *ACM Trans. on Comp. Syst.* **2**(4), 278–288 (1984).
19.   R. Stewart, *et al.,* Stream Control Transmission Protocol (SCTP) dynamic address reconfiguration, *draft-ietf-tsvwg-addip-sctp-08.txt,* Sept. 2003 (work in progress).
20.   M. Riegel and M. Tuexen, Mobile SCTP, *draft-riegel-tuexen-mobile-sctp-03.txt,* Aug. 2003 (work in progress).
21.   S.J. Koh, *et al.,* Mobile SCTP for transport layer mobility, *draft-sjkoh-sctp-mobility-03.txt,* Feb. 2004 (work in progress).
22.   M. Buddhikot, *et al.,* Integration of 802.11 and third-generation wireless data networks, in *Proc. IEEE INFOCOM,* San Francisco, CA, Apr. 2003.
23.   L. Ma, F. Yu, V.C.M. Leung and T. Randhawa, A new method to support UMTS/WLAN vertical handover using SCTP, *IEEE Wireless Commun.,* (2004) (in press).
24.   L. Ma, *A new method to support UMTS/WLAN vertical handover using SCTP,* MASc thesis, University of British Columbia, 2004.
25.   http://pel.cis.udel.ed.
26.   T. Dierks and C. Allen, The TLS protocol version 1.0, *IETF RFC 2246,* Jan. 1999.
27.   M. Abadi and R. Needham, Prudent engineering practice for cryptographic protocols, *IEEE Trans. on Software Engineering* **22**(1), 6–15 (1996).
28.   E.K. Kwon, Y.G. Cho, and K.J. Chae, Integrated transport layer security: End-to-end security model between WTLS and TLS, in *Proc. IEEE ICOIN,* Beppu, Japan, Jan. 2001.
29.   Y. Song, V.C.M. Leung and K. Beznosov, Supporting end-to-end security across proxies with multiple-channel SSL, in *Proc. IFIP SEC,* Tolouse, France, Aug. 2004.

Chapter 15

DEVELOPMENT AND FUTURE APPLICATIONS OF SATELLITE COMMUNICATIONS

ERICH LUTZ, HERMANN BISCHL, HARALD ERNST, FLORIAN DAVID, MATTHIAS HOLZBOCK, AXEL JAHN, MARKUS WERNER
DLR – Deutsches Zentrum für Luft- und Raumfahrt, Oberpfaffenhofen, D-82234 Wessling, Germany

Abstract: The paper describes some essential trends in the development of satellite communications and introduces promising future applications. Important developments are the set up of standards based on digital video broadcast (DVB) and the application of on-board switching taking into account Quality of Service (QoS). Other developments are the application of multiprotocol label switching (MPLS) to satellite networks and the usage of optical inter-satellite and inter-platform links. Seamless interworking with terrestrial wireless access networks and with terrestrial core networks is of paramount importance for the success of satellite networks. Future applications of satellite communications include aeronautical satellite communications, satellite-based air traffic management and personalised land mobile radio broadcasting.

1. INTRODUCTION

Starting in 1957 with the launch of Sputnik, satellite communications experienced a long successful history and will further develop into the future. Satellite communications has a number of features:

- In general, satellites have an inherent broadcast characteristic and can cover large areas. They are useful for TV and radio broadcast, as well as for (reliable) data multicast (content distribution). E.g., Web content can be distributed to ISPs and caches at the Internet edge.
- Similarly, satellites are advantageous for building wide area (closed) networks for distributed company and institution branches (VSAT networks). Satellite on-board switching can specifically enhance the efficiency of such networks.

- Because of their large coverage area, satellites can effectively provide rural areas and developing countries with information and communications services (bridging the digital divide).
- Similarly, satellites are useful for providing wide area mobile users (land, maritime, aeronautical) with broadcast and two-way communications. Moreover, they can support fleet and air traffic management.

Facing stiff competition from terrestrial communications systems, it is of crucial importance for the commercial success of satellite systems to apply them to communications services and scenarios where they are most useful and advantageous.

2.   SYSTEM TRENDS

2.1   Geostationary Satellite Systems

Geostationary satellites are placed into an equatorial orbit at 35786 km where they appear stationary with regard to an observer on the Earth's surface[1]. This is an advantage e.g. for DTH TV reception, allowing fixed satellite dishes. In spite of the high signal attenuation caused by the large distance, GEO satellites are frequently used for mobile services (e.g. Inmarsat). Modern GEOs compensate for this effect by employing large antennas (e.g. 12 m reflectors for Thuraya and ACeS). Such antennas, however, produce narrow spot beams, a large number of which must be used to fill the coverage area. E.g., Thuraya uses about 250 spot beams generated by a multi-feed reflector antenna.

2.2   Non-geostationary Satellite Systems

To avoid the large distance to GEO satellites, system concepts using non-geostationary satellites in low or medium earth orbits (LEO, MEO) have been developed in the nineties. This approach allows smaller satellites and significantly reduces signal delay, which is crucial for conversational services. While modern GEO satellites have high mass (typically 5 tons) and high power (typ. 12 kW DC), LEO satellites can be much smaller and cheaper. In order to provide a required radio cell pattern on the Earth's surface, a LEO satellite needs to implement much less spot beams compared to a GEO. On the other hand, in order to achieve continuous coverage of an area, many LEO satellites must be deployed, automatically leading to a global coverage. Iridium and Globalstar are such revolutionary systems, which have not been commercially successful due to the quick deployment of terrestrial mobile networks.

2.3 High Altitude Platform Systems

New concepts foresee the usage of high altitude platforms as a possible substitute for communications satellites. Such platforms flying geostationary in altitudes around 20 km would combine the advantages of GEO satellites (geostationary position) and LEO satellites (short distance). They are intended for providing broadband services as well as mobile services (UMTS and beyond). Probably the most severe obstacle for implementing such systems are high wind speeds that have to be overcome by the platforms. Platform concepts include heliostatic platforms, as well as high altitude long operation aircraft.

3. TECHNOLOGY DEVELOPMENT

3.1 Standards based on Digital Video Broadcast (DVB)

The Digital Video Broadcasting – Satellite (DVB-S) and DVB – Return Channel via Satellite (DVB-RCS) standards[2] have been developed to bring digital television to homes, but due to the use of MPEG2 packets and ATM cells as data containers they are also designed to support IP based data services via satellite. The DVB standards have been recently evolved to their next generation. E.g., in order to achieve better bandwidth efficiency and robustness against rain fading, adaptive coding and modulation for point-to-point connections has been introduced in the new DVB-S2 standard.

3.2 On-Board Switching

Future broadband satellite networks incorporate on-board switching and are seamlessly integrated with terrestrial networks, in particular with the Internet. It is therefore reasonable to transport IP-based traffic also in the broadband satellite networks. On-board switching offers considerable improvements in spectrum sharing and QoS provisioning for future broadband satellite networks[3]. In contrast to most of the existing satellite systems acting as "bent pipes", onboard switching reduces delay and saves bandwidth and power by accomplishing connections with only one satellite hop instead of two, Fig. 1.

On-board switching can be performed in different protocol layers. Physical layer switching means that a TDM matrix is used to switch physical layer bursts. This requires synchronisation of the on-board switch with the terminals on ground. Layer 2 switching is based on labels of the layer 2 packets. In case of DVB-S/RCS this could be MPEG2 switching based on the DVB PID (packet identifier) values. An example for layer 3 switching is IP switching, which requires de-encapsulation/re-encapsulation of the IP datagrams on board the satellite.

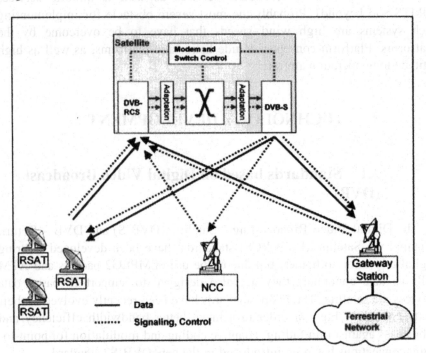

Figure 1   DVB-S/DVB-RCS broadband satellite system with on-board switching. RSAT = Regenerative Satellite Access Terminal

Examples of broadband system concepts with on-board switching are the Astrolink concept envisaging an on-board ATM switch, the EuroSkyWay concept with on-board switching of ESW cells and support of DVB, IP, and ATM, and the AmerHis initiative, combining DVB-S and DVB-RCS into a regenerative multi-spot beam satellite system.

3.3 QoS-aware Onboard Switching

Future satellite systems should support IP-based services with Quality of Service (QoS) techniques like IntServ and DiffServ in order to meet the network requirements of broadband multimedia applications. In a fully meshed DVB-S and DVB-RCS compliant satellite system, the on-board switch is the keystone of traffic transfer: it is supposed to receive and centralize the total traffic generated by the whole community of users (both, signaling and data traffic coming from satellite terminals, satellite gateways, and Network Control Center), to process this traffic and to route it further towards its final destination. This set of operations should be performed taking into account QoS parameters associated with each individual traffic flow.

Corresponding to the type of packets transported (IP datagrams with DiffServ or IntServ, MPLS packets, or MPEG2 frames), the on-board switch should be able to support either per-flow or per-category QoS mechanisms.

3.4 Multiprotocol Label Switching (MPLS)

Non-geostationary satellite constellations with inter-satellite links (ISL) are a challenge for networking, due to their continuously changing topology. In order to make maximal usage of the networks' capacity, specific attention has to be paid to routing and traffic engineering. Multiprotocol label switching as underlying protocol is an interesting candidate for this task since it offers many possibilities to exert influence on traffic flows and supports today's dominating Internet protocol traffic very well[4]. In MPLS the task of routing definition is allocated to label edge routers at the network edge, Fig. 2. Here, various constraints such as QoS or traffic engineering can be taken into account. The label switching routers within the MPLS cloud just have to perform simple and fast label switching (label swapping).

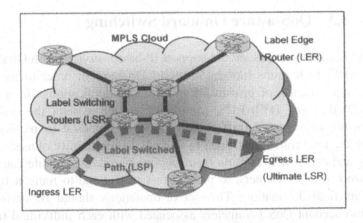

Figure 2 MPLS network architecture

MPLS has already been proposed for GEO systems with multiple spot beams, mainly as a means to reduce the complexity of on-board switching. In LEO constellations with ISLs a potentially high traffic load must be switched between ISLs. Here, MPLS can be used for traffic engineering taking account of the time-varying constellation topology.

3.5 Optical Inter-Satellite and Inter-Platform Links

To cope with the increasing demand on bandwidth, optical free-space transmission appears to be a promising alternative to conventional microwave links for interconnecting satellites or high altitude platforms (HAP). The short optical wavelengths in the range of microns allow for very narrow beams and hence enable a reduction of weight, size, and power consumption of ISL terminals. The bandwidth provided by optical terminals is almost unlimited. The DLR laser communications terminal (LCT) for optical inter-satellite links (OISL), developed by Tesat under DLR funding and with fundamental DLR research contributions, is capable of transmitting several Gbps using advanced coherent technology. It will be verified in space in 2006 as secondary payload on-board the LEO satellite TerraSAR-X. Wavelength division multiplexing (WDM) will allow for a further increase in data rates in future systems.

Future applications of optical terminals may include GEO to GEO satellite links, geostationary data relay satellites, optical back-bone networks, optical downlinks from earth observation satellites to dedicated optical ground station networks, HAP networks, and high-speed long-haul connections between unmanned aerial vehicles (UAVs) or from UAV to GEO relays, Fig. 3.

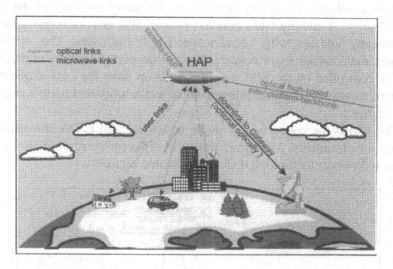

Figure 3 High Altitude Platform scenario including optical free space communication links

While OISLs are only perturbed by micro-vibrations of the satellite, inter-platform links between HAPs suffer from significant motion and vibration of the platform as well as from severe atmospheric disturbances[5]. The turbulent atmosphere leads to intensity fluctuations and wave-front distortions causing very deep and extremely slow fading. Current research activities deal with the modelling of the atmospheric transmission channel, the optimisation of transmission schemes for this particular channel, and the mitigation of atmospheric disturbances e.g. by the use of diversity schemes.

3.6 IP Mobility in Heterogeneous Satellite – Terrestrial Wireless Networks

In future mobile communications users expect to have access to Internet and PSTN from everywhere, using their own personalised devices via different access networks being available in different scenarios. This approach calls for transparent interworking of the access networks. Collectively mobile networks, e.g. formed by passengers in an aircraft[6] are a particular case.

The concept of user mobility with seamless handover is based on the idea that the IP addresses of the user terminals do not change, while the Segment Interworking Service Integrator (SISI) is changing the active access network segment, Fig. 4. Hence, all IP based applications do not notice any change, apart from parameters like data rate or delay, and no user interaction is required when passing the coverage borders of access networks.

Such a solution is mainly realised by a routing function of the SISI functionality. A routing table handled by the SISI and its home agent is permanently updated with segment-specific IP addresses. The different segment radio modems are connected to the network interfaces of the SISI and are controlled by the SISI with a call set-up layer for each segment. A segment monitoring module informs about availability and performance of each segment.

This allows the connection of single nodes and also LANs consisting of several user terminals. Moreover, traffic of different user terminals can be routed independently through different available segments.

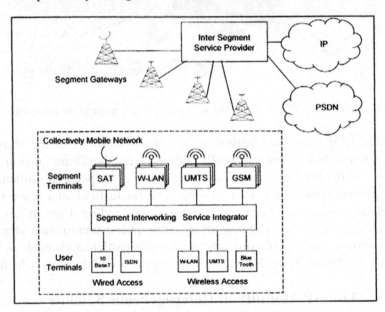

Figure 4 Heterogeneous wireless network scenario with a collectively mobile network and different wireless access networks (segments).

4.   FUTURE APPLICATIONS

4.1 Aeronautical Satellite Communications

Providing broadband communications to aircraft passengers (AirCom) via satellite is one of the major application trends in the satellite field[7]. The introduction of new global broadband systems by Inmarsat to provide broadband access in the sky or the collaboration between Lufthansa and Connexion by Boeing in the FlyNet project to provide Internet connectivity on-board aircraft reflect a huge market potential. While in the past aircraft operators could only rely on low-rate communication service with only a few kilobits per second used for voice and for low rate communication, now links with data rates of several hundred kbps and up to some Mbps are available to allow a new variety of services and applications.

Two paths are followed for the evolution of aeronautical services:

- improving the aircraft cabin technology by a user-friendly environment for personal and multimedia communications, and
- improving the space-segment technology to allow higher bit rate services.

In the space segment, several parallel activities target the aeronautical market. Inmarsat will increase the data rate of its aeronautical services and the system capacity using about 200 spot beams with its 4th generation satellites. In the B-GAN™ system that will be deployed in 2005, the available data rate ranges up to 432 kb/s.

New agile aircraft antennas are needed to provide high-gain characteristics at Ku band and above. The Connexion™ by Boeing system is one of the potential candidates using such Ku-band satellites for two-way communications and the European airlines British Airways and Lufthansa have early demonstrated and tested this approach. Moreover, Lufthansa and Japan Airlines are starting the service roll-out across their long-haul fleet with such equipment.

Currently, the European project *WirelessCabin* (www.wirelesscabin.com) aims at providing aircraft passengers and crew members with heterogeneous wireless access solutions for in-flight entertainment, Internet access and mobile/personal communications. Aircraft passengers will be offered the same wireless services for personal and multimedia communications as they are on ground. The most important wireless access technologies are GSM, UMTS, W-LAN IEEE 802.11, and Bluetooth™.

The *WirelessCabin* system architecture comprises:

- Several wireless access segments in the aircraft cabin, namely a wireless LAN according to IEEE 802.11b standard for IP services, a GSM/UMTS pico-cell for personal and data communications, and Bluetooth™ 1.1, as well as a standard wired IP LAN, Fig. 5.
- A satellite segment for interconnection of the cabin with the terrestrial telecom networks. The different cabin services must be integrated and interconnected using a *service integrator,* which allows the separation

and transportation of the services over a single or several satellite bearers.

- An aircom service provider segment supporting the integrated cabin services. The aircom provider segment provides the interconnection to the terrestrial personal and data networks as well as the Internet backbone, Fig. 6. For the GSM/UMTS cabin service, part of the core network must be available.

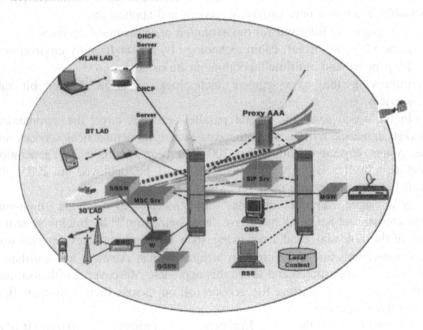

Figure 5 System architecture of the *WirelessCabin* cabin segment

A wide bouquet of *WirelessCabin* services will be demonstrated in flight with an Airbus aircraft in summer 2004, to be succeeded by a pilot phase with a major airline. The operational service roll-out could start in a few years.

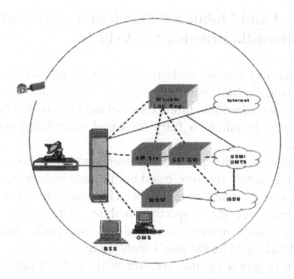

Figure 6 System architecture of the *WirelessCabin* ground segment

4.2 Air Traffic Management

Advanced satellite-based aeronautical passenger communications, as soon to be provided from *Wireless Cabin,* inherently bears the potential to boost some improvements in air traffic control (ATC) communications, just because of the sheer appeal of available infrastructure, equipment, and capacity. The take-up of usage for ATC purposes will start from less safety-critical applications such as crew communications, general airline operational control, and bi-directional data transfer of supporting air traffic management (ATM) information between cockpit and control centers.

With *WirelessCabin* and a global or regional satellite aeronautical passenger communication infrastructure in place both, pilots and controllers can benefit from separate high-quality voice channels that can be reserved and prioritized within the permanent traffic multiplex in both directions between aircraft and ground. Such service would be available in the whole coverage area of the satellite system. In addition to voice communication, bi-directional data transfer of relevant ATM information between cockpit and control centers can add extraordinary value to the quality of ATC operations; this may range from simply securing inaudible or fault-prone voice commands, up to a permanent real-time update of a complete picture of the surrounding airspace.

4.3   Land Mobile Satellite Radio – Personalised Multimedia Broadcast to Vehicles

Related to the introduction of digital radio broadcast and the delivery of content as multimedia files, new opportunities for satellites arise in the area of land mobile radio broadcast. In the frame of the ESA project "Mobile Ku-band Receive-only Terminal" a fully file-based radio broadcast concept is being designed.

In this concept, an individual entertainment and information programme is compiled at the mobile receiver, based on files that have been successfully received or previously stored in the cache. This programme can be tailored to the preferences of the respective mobile user. In particular, the programme can contain news services and real-time traffic information in intervals and detail based on the user's wishes, Fig. 7.

In the ESA project a satellite receiver will be developed and the file-based radio concept will be demonstrated via the Astra 1A satellite.

A major problem in land mobile satellite transmission is the presence of time-varying signal shadowing due to obstacles in the transmission path. The file-based radio transmission is able to overcome this problem by applying a suitable error control protocol. In addition to channel coding at the physical layer, this protocol concept also applies transport layer coding: Each programme file is divided into a number of IP packets, and specific redundancy packets are added, allowing reconstructing of the file in case of packet losses. The packets constituting different programme files can be extensively interleaved and spread over a long time interval, in order to overcome signal shadowing events[8]. Moreover, adding a return link, this concept enables reliable distribution of information, based on hybrid ARQ type II multicast retransmission techniques.

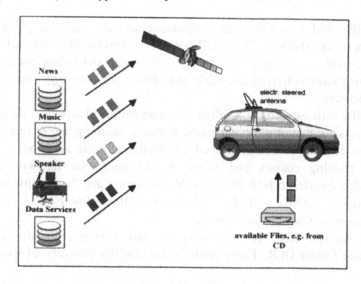

Figure 7 File-based personal radio concept

5. THE EUROPEAN SATELLITE COMMUNICATIONS NETWORK OF EXCELLENCE "SATNEX"

SatNEx is a Network of Excellence in the area of satellite communications and is financed by the European Commission within its 6th framework programme. SatNEx will be active in the field of mobile and broadband satellite communications. According to the expectations of the EC, the SatNEx NoE will concentrate on a Joint Programme of Activities, comprising Integrating Activities, Jointly Executed Research, and activities dedicated to the Spreading of Excellence.

The primary goal of the SatNEx NoE is to achieve long-lasting integration of the European research in satellite communications and to develop a common knowledge base. Through co-operation of excellent universities and research organisations with outstanding expertise in satellite communications, SatNEx will build a European virtual centre of excellence in satellite communications and contribute to the realisation of the European Research Area.

SatNEx includes an advisory board incorporating representatives of European space industry, satellite service providers, and standardisation and regulation organisations. SatNEx will collaborate with these players and put together a critical mass of resources and expertise needed for making Europe a world force in the field of satellite communications.

SatNEx will disseminate the knowledge and the expertise generated to researchers and students. They will benefit by moving between institutions of the network and using special facilities. Training and lecture material will be electronically delivered and made available to all researchers and students in the network.

SatNEx will establish a satellite communication platform to be used by all partners as a means for integrated research, teaching and training. Also, common test beds will be developed, integrating existing local tools.

Via training courses and skills development, SatNEx will achieve knowledge transfer to both the research community and the satellite industry and business. SatNEx will also look to the creation of new business via the exploitation of research knowledge.

The SatNEx NoE is co-coordinated and managed by the German Aerospace Center DLR. Participants of the SatNEx Network of Excellence are:

- German Aerospace Center (DLR) Germany
- Aristotle University of Thessaloniki Greece
- University of Bradford United Kingdom
- Budapest University of Technology and Economics Hungary
- Centre National d'Etudes Spatiales France
- Consorzio Nazionale Interuniversitario per le Telecommunicazioni Italy
- Fraunhofer Gesellschaft zur Förderung der Angewandten Forschung e.V. Germany
- Groupe des Ecoles des Télécommunications France
- Institute of Communication and Computer Systems of NTUA Greece
- National Observatory of Athens Greece
- Istituto di Scienze e Tecnologia dell'Informazione "Alessandro Faedo" Italy
- Jozef Stefan Institute Slovenia
- Rheinisch-Westfälische Technische Hochschule Aachen Germany
- Office National d'Etudes et de Recherches Aérospatiales France
- Graz University of Technology Austria
- Universidad Carlos III de Madrid Spain
- The University of Surrey United Kingdom
- Department of Engineering, University of Aberdeen United Kingdom
- DEIS - University of Bologna Italy
- Università Degli Studi di Roma "Tor Vergata" Italy
- Universidad De Vigo Spain.

The SatNEx Network will run over two years (2004/2005). Its 21 partner organisations from 9 European countries contribute with 182 researchers and will have an overall funding volume of approx. 4 million Euro.

REFERENCES

1. E. Lutz, M. Werner and A. Jahn, Satellite Systems for Personal and Broadband Communications, Springer, Berlin Heidelberg, 2000.
2. ETSI EN 301 790 V1.3.1 (2003-03). European Standard (Telecommunications series), Digital Video Broadcasting (DVB); Interaction channel for satellite distribution systems.
3. N. Courville, H. Bischl and J. Zeng, "Critical issues of on-board switching in satellite DVB-S/RCS networks", to be published.
4. A. Donner, M. Berioli and M. Werner, "MPLS-based satellite constellation networks", IEEE J. SAC, vol. 22, May 2004.
5. F. David, D. Giggenbach, H. Henniger, J. Horwath, R. Landrock and N. Perlot: "Design Considerations for Optical Inter-HAP Links", in Proceedings of the 22nd Int. Communications Satellite Systems Conf. (ICSSC), AIAA, Monterey, CA, May 2004.
6. M. Holzbock, "IP based user mobility in heterogeneous wireless satellite-terrestrial networks", Wireless Personal Communication, Special Issue on Broadband Mobile Terrestrial-Satellite Integrated Systems, vol. 24, no. 2, pp. 219-232, 2003.
7. A. Jahn, M. Holzbock et al, "Evolution of aeronautical communications for personal and multimedia services", IEEE Comm. Mag., July 2003.
8. H. Ernst, L. Sartorello and S. Scalise, "Transport layer coding for the land mobile satellite channel", VTC 2004-Spring, Milano, Italy, May 2004.

Chapter 16

ROLE OF SATELLITES IN MOBILE/WIRELESS SYSTEMS

B G EVANS
CCSR, University of Surrey, Guildford, Surrey GU2 7XH, United Kingdom

Abstract   It has long been a conjecture as to whether satellites have a role in Mobile Communications other than for niche areas such as sea and aero coverage. This paper attempts to answer this question by looking at the history of mobile satellite communications systems and the current developments towards multimedia, broadband, content and integrated systems. It projects into what is called the B3G era to show how satellites can play a useful role in cooperation with terrestrial systems.

1.   DON'T MENTION SATELLITES!

Whether it is in conferences or in standards bodies, the dominance of the terrestrial players dictates that satellites always get a pretty rough ride –even if they are allowed on stage in the first place. This has always struck me as very strange as fixed satellite communications has always been a part of most large 'terrestrial' telecom operators' portfolio and one that has produced considerable revenue. In the mobile area, INMARSAT [1] has conducted a profitable business for the last twenty years or so and has not impinged at all on the terrestrial mobile operators business. So why is 'Space' seen as the 'opposition' and a threat to telecom operators? It could be ignorance, or an imagined threat of the unknown. It may be the relative isolation of satellite systems and the fact that they have emerged with different operators that control the major resource of the satellite in space. Whatever the reason, it is completely unfounded as satellites are just another delivery mechanism for telecom services and their use will depend on economic arguments just like the choice of cable, fibre or radio.

For unicast type services such as speech and low data rate services which have predominated the past, satellites are only economically viable in areas such as the oceans, air or sparsely populated landmarks. Their limited power and allocated frequency spectrum makes them poor contenders on a point-to-point basis with other delivery mechanisms. The advantages of satellites have long been known to be:

- Wide coverage – broadcast;
- Speed to deliver new services.

This explains why they have been pre-eminent in delivery of broadcast services -80% of digital television is received via satellite in Europe. Also why the major traffic carried by fixed service satellites is now INTERNET for ISP's.

The adversaries have cited delay and high error rates as being major detractors for satellites. With GEO satellites, delay can cause problems for speech services if terrestrial operators cannot control their echo problems but with modern echo-cancellers and properly set-up networks, this is not a problem today. Satellite channels are subject to rain fades for fixed services but using modern modulation and coding schemes, deliver services at bit error rate (BER) of better than 10^{-10}.

So what of the use of satellites for mobile services? Again for speech and low data rate services, the twin limitations of satellite power and radio spectrum do not allow them to compete in efficiency terms with terrestrial cellular. Thus the successes have been in the niche areas of sea and aero which have been covered by INMARSAT. In mobility terms, they are also making an impact in radio and television broadcast to those on the move and as we move to broadband possibly to content provision. So we return again to the two major satellite advantages given above –broadcast and speed of delivery and this is where they are likely to make maximum impact in the future. However before we explore this, let's look at the lessons delivered by history.

2.   A SHORT HISTORY OF MOBILE SATELLITE COMMUNICATIONS

We show in Table 1 some of the key ideas and the major mobile satellite systems (MSS) that have resulted from them. It is interesting to note that the major mobile satellite operator, INMARSAT, came into existence at around the same time as the first cellular operators providing 1G analogue services. In its initial period, INMARSAT provided speech and low data rate services mainly to the maritime market of larger ships in L band using global beam coverage satellites. In 1990/1, INMARSAT added aeronautical services to passenger aircraft and to some land vehicles with the introduction of spotbeam higher power satellites. This was followed in 1997/8 with world wide spotbeam operation in the MSS and the introduction of paging, navigation, and higher rate digital to desktop sized terminals. INMARSAT have concentrated on the use of GEO satellites and in the mid 90's several regional GEO systems emerged in competition e.g. OMNITRACS, EUTELTRACS, AMSC and OPTUS concentrating on land vehicles and using both L and Ku bands. These were only moderately successfully whilst INMARSAT built its customer base to around 200,000. The major research in the late 80's and early 90's was in non-GEO constellations and this saw proposals for LEO and MEO satellite systems. Of these, Iridium and Globalstar came into services, but too late to compete with the spread of terrestrial GSM, and on business, rather than technological grounds went into Chapter 11 bankruptcy by the early 2000's. The lesson to learn was that constellations were too expensive, at up to $10B, to deploy unless markets had large initial growths to pay back. Both systems are in existence today but with a fewer customers than initially predicted. (Orbcomm, a little LEO provider mainly to fixed terminals suffered a similar fate). ICO, the proposed MEO system got as far as launching one satellite before realising also that the business case was not there.

Table 1 Mobile Satellite Developments

cellular		Research Ideas	Operational systems (mobile)	(broadcast)
1G	1970's	Mobile sat. expts ATS-6	Inmarsat formed	
	1980's	Non-GEO mobile cellular architecture proposed University of Surrey	Inmarsat operates – maritime	
2G GSM	1990's	Motorola announce Iridium system Little LEO Orcom system proposed - Teledesic announce non-GEO fixed systems - Globalstar/ ICO proposed - Super GEO's announced Agrani/ Apmt/ Aces/ Thuraya	Inmarsat operates –land/ aero Regionals: Omnitracs, Euteltracs, Amsc, Optus - Inmarsat Sats-spots - Iridium operational - Orbcom operational - Globalstar operational	(world space radio)
3G IMT-2000	2000's	- Integrates S/T UMTS for content proposed Satin EU project –University of Surrey - DVB-S2 standard	- Iridium/ Globalstar/ Orbcoms - ch11 continue operate restricted modes - Thuraya operational - Inmarsat IV's -100's spots and dsp processor	(Xm, SIRIUS, DARS) (MBSAT)

In the mid 1990's, larger so-called super GEO [2] satellites were proposed, these being around 5kW with 100-200 spots rather than the earlier generation of GEO's 3-4kW with 5-10 spots. Several such systems were proposed, but the one that has reached market in the early 2000's is Thuraya [3] [4] (which is based on ETSI GMR-1 standard [5] - Fig. 1) providing GSM and GPRS like services covering Asia and much of Europe (Fig. 2). Although it is early days this super GEO seems to be successful, finding a niche with travellers, trucks and in areas where terrestrial mobile is expensive to deploy. Meanwhile INMARSAT is providing its own super GEO's-INMARSAT IV to take existing digital services from 64kb/s up to 432kb/s – from the Global Area Network (GAN) to the broadband GAN (BGAN) [6], which will be launched in 2004. Despite the move by terrestrial mobile operators to CDMA, INMARSAT has continued to develop its proprietary TDMA system but deliver 3G equivalent packet services

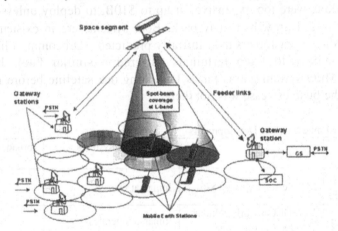

Fig. 1: GEO Mobile Radio (GMR) System Architecture [5]

Fig. 2. Thuraya satellite and terminal

So the lessons that have been learnt are that with so called basic unicast services, satellites can only economically provide niche services to areas inaccessible to cellular. Constellations have proved too expensive to compete with GEO's or cellular systems and there is now a return to GEO.

We note in the above that mobile satellites have concentrated in traditional services and due to their expense in space segment provision to niche markets. Now can this change?

3. CONTENT DELIVERY AND INTEGRATION

Of the lessons learnt, there are three that predominate:

i) Not to compete but collaborate with cellular;
ii) Use the wide coverage broadcast attribute of satellites;
iii) Select the service appropriate to the delivery mechanism.

It was this realisation that drove us to propose in an EU FP5 project (SATIN) [7][8] a new way forward for mobile satellites in the form of an integrated satellite/terrestrial-UMTS system. At around this time the transition to 3G-IMT-2000 systems was based upon new multimedia services and a predominant sector of such services was content delivery, or push type services. The rapid advance in memory devices and their reduction in cost meant that personal terminals could store and play back large content volumes. Services of this type are different from those basic services catered for by cellular systems to date. They are more multimedia broadcast and multicast services (MBMS) [9][10][11] in nature which well suits them to satellite delivery. It also transpires that such services are not so efficiently delivered in a cellular environment where they severely reduce the capacity for basic services. This all points to an integrated system between satellite and cellular in which the services are divided to the delivery mechanism that best suits them. The architecture of such a system is shown in Fig. 3 from SATIN which demonstrated the feasibility of the system and a follow-on project MODIS [12][13] has later this year will demonstrate the integrated system using the Monaco 3G network. An FP6 project MAESTRO [14] is now in place which will take these 'proofs of concepts' to a fully operational satellite digital multimedia broadcast system (SDMB-Alcatel) for operation in 2007/8. The architecture is characterised by gap-fillers or Intermediate Module Repeaters (IMR's) located at some 3G base stations [15], which broadcast the MBMS signals terrestrially in the adjacent MSS band to allow in-building and urban presentation.

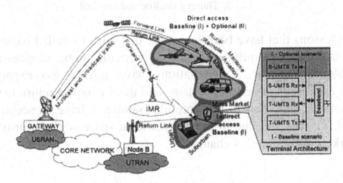

Fig. 3. SATIN Reference Architecture

A similar concept has been adopted for the MBSAT [16] system in Japan and Korea where the service driver is mobile television rather than content. Fig. 4 shows the system, which is to become operational this year.

The Digital Audio Broadcasting (DAB) systems via satellite –S-DAB (DARS in the US) should also be mentioned in this context as radio channels are a further example of content. The idea has been around since 1990 when CD Radio first filed in the US. Several systems have been proposed since the S-DAB standards were produced with WORLDSPACE [2][17] in the mid 1990's being perhaps the leading contender with the satellites to cover Asia, Caribbean and the Americas.

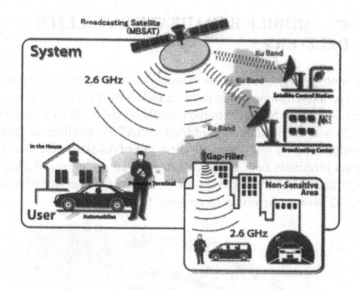

Fig. 4: Japanese Mobile Broadcast Satellite System (MBSAT)

The terrestrial equivalent T-DAB, has not spread widely with the limited UK network being perhaps the best developed. In the US in early 2000's two commercial systems have become operational; Xm Radio –using GEO satellites and SIRIUS radio using HEO's [2][17]. Both systems use terrestrial gap-fillers in a similar way to that proposed by SDMB and MBSAT. The use of HEO satellites are interesting in that they achieve improved coverage in the urban area and reduce the number of gap-fillers required. Currently Xm has around 1 million and SIRIUS 250,000 customers in the US.

S-DAB and MBSAT are similar in that they are content driven –radio in the former and television in the latter, but each could provide both services. SDMB is multimedia content driven but could provide radio and television and is the only system that is specifically aimed at integration with cellular. It is this latter concept which completes the picture and opens up a whole new future for satellites. Each of these systems has regulatory issues to overcome in various regions of the world and this could well be the determinant as to which succeeds in the future. The services covered in this section might be described as 'mid-band' but can satellites progress from them to broadband delivery in a mobile scenario? (An approximate definition of mid-band would be 100's kb/s to a few Mb/s).

4. MOBILE BROADBAND SATELLITE DELIVERY

In the mobile sense, broadband is much less than would be defined in a terrestrial fixed environment. For instance INMARSAT will be moving from the current 64kb/s GAN system to the broadband GAN (BGAN) at up to 432 kb/s (Fig. 5) with the introduction of INMARSAT IV satellites in 2004 (Fig. 6). These higher power satellites will allow use of A4 sized notebook and A5 sized pocket terminals both of which can be used as 'plug and play' with laptop PC's and PDA's respectively. In this respect the service offering can be akin to 3G network offerings to personal terminals.

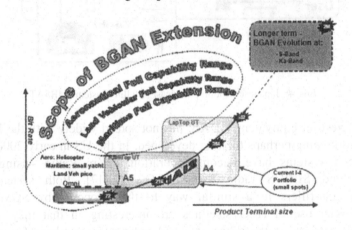

Fig. 5: BGAN Evolution

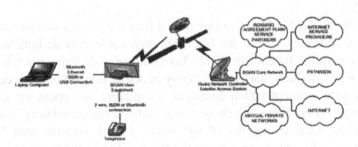

Fig. 6: INMARSAT BGAN Architecture

Another major market is for passenger vehicles –aircraft, ships and trains/coaches where customers are perhaps more likely to use broadband services. Besides INMARSAT, Connexions by Boeing [18][19] began operating broadband links to airplanes in 2002 and is now pursuing the maritime operators market. The technology here has been more akin to the VSAT model with local in vehicle distribution. Connexions has already installed terminals with a number of airlines. VSAT systems started in the offshore oil business but have rapidly expanded to cruise liners and deep-sea ferry operators using Ku band and provide commercial, engineering and navigation services to passengers and crew. A number of satellite operators carry such services. Extensions can also be made to land vehicles and EU FP6 projects DRIVE/OVERDRIVE [20] have researched the coach/car markets and FIFTH the high-speed train market.

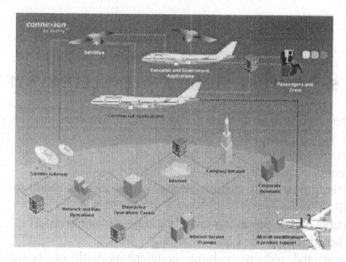

Fig. 7: Connexion by Boeing Broadband satellite network

The above schemes still suffer from the poor efficiency of use of the satellite capacity which makes them expensive. A solution to this may be around the corner with the introduction of the new DVB-S2 standard this year [21]. Principally aimed at fixed systems it incorporates adaptive coding and modulation (ACM) schemes which when operated in connection with the return satellite channel RCS allows transmission parameters to be optimised for each individual connection dependent on path conditions. A range of PSK, APSK modulation schemes and LDPC codes provide a packet by packet optimisation to meet the adverse changing channel conditions. The new standard allows a range of data inputs including IP. Combining the DVB-S2 ACM scheme with multi-spot Ka band satellites and DVB-RCS return link current satellite capacity can be increased by a factor of 10. The next step is to introduce mobility into the standard which will then enable it to be used for broadband mobile multimedia connections as mentioned above.

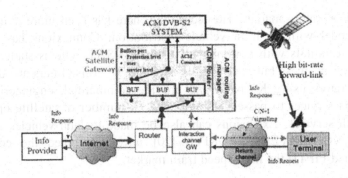

Fig. 8: Example of IP services using a DVB-S2 ACM link [21]

5. FUTURE MOBILE SATELLITE SYSTEMS

For the medium term future, GEO MSS satellites are expected to dominate with higher power, increased number of beams and processed bandwidth. Key technologies will be:

- Larger deployable reflectors up to tens or meters
- Higher power multi-beam antenna with adaptive beam shaping via DSP
- Scalable digital processors enabling improved connectivity
- On board regeneration and switching
- Lighter and reduced volume components with on board wireless connections
- Short range adhoc satellite clusters

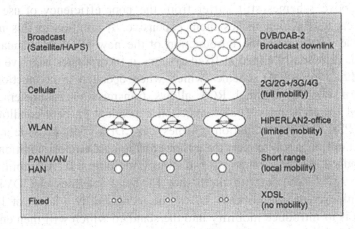

Fig. 9: Vertical Hierarchical Networks

Such satellites will fit into the unified view of the 4G mobile vision produced by WWRF [22] as shown in Fig. 9 at the top layer. However this vision also provides for global roaming across all layers and points the way to full integration of terrestrial and satellite systems at terminal and network levels with handover across all the networks. The terminal accesses any network and service to maximise the cost and efficiency. Such a scenario will only be possible if operators embrace the totality of the delivery mechanism and network providers (both satellite and terrestrial) embrace common standards. A key enabler to this scenario will also be SDR (software defined radio), which will enable terminals to automatically reconfigure to different networks and middleware which will automatically configure the most cost effective connections. We see the beginning of such an approach with the S-DMB system in integrating multicast/ broadcast via satellite with other 3G services terrestrially (Fig. 10). But this is just a small step. To what extent this integration will penetrate the vertical layer structure is a matter of conjecture. As well as the cellular layer already mentioned there are experiments from satellite to the WLAN or Wi-Fi layer in the FP6 project TWISTER. Looking further ahead one can also envisage a connection to the PAN layer particularly in respect of location of the entities and as part of a much enhanced sensor network. Also Alcatel's 'ADSL in the sky' system would provide the integration from the top to the bottom of layered stack in Fig 9.

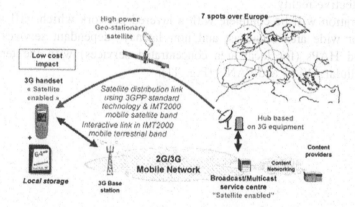

Fig. 10: Satellite Digital Multimedia Broadcast concept

In the medium-term the drive will be broadband for non-broadcasting applications with particular emphasis on passenger vehicles – planes/ ships and road vehicles. Broadcast and multicast applications have already been mentioned and will develop in both non-real-time (e.g. S-DMB) and real-time (MBSAT/ DAB) areas.

Integration is the single most important feature and challenges exist in the efficient integration of multicast and unicast capabilities, improving the cost per bit and capacity, optimising physical/ MAC layers between the systems and all this in a common IP QoS environment. Besides these technical challenges, there are the cultural challenges for the operators and network providers who must come together to develop new systems within the standards arena rather than go on ploughing their own furrows. The latter is possibly the biggest challenge!

Finally what of the longer term future – 2020. In the last 20 years, satellites have become 1000 times more powerful and similar times more cost effective. There does not seem anything in the laws of physics to prevent this trend continuing. There will be no long term needs for gap fillers as satellite power will be sufficient to provide in building and urban coverage. Constellations will make a comeback with HEO's providing the efficiency for broadcast and LEO/ MEO's for solving the latency problem.

Adaptive air-interface technologies and reconfiguration will remove the current limitations of the satellite channel.

Perhaps the most radical, and the enabler of the resurgence of constellations, will be the emergence of smaller and cheaper satellites. These will enable whole LEO constellations of hundreds of satellites complete with onboard IP routers to be deployed for 10's M$ rather than 10's B$. The original Teledesic concept of a complete IP network in the sky will become a cost effective reality.

Integration will be complete with a layered network which will include GEO (for wide area broadcast and non latency -dependant services), non-GEO and HAPs (for local area concentrated services) with the terrestrial based cellular, Wi-Fi and PAN's (Fig. 11).

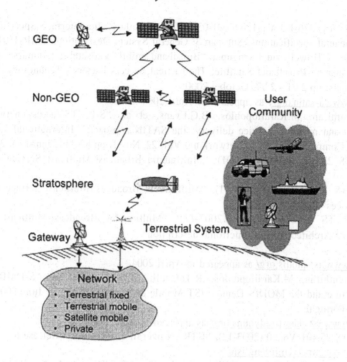

Fig. 11: Integrated terrestrial, satellite, HAPS system

These are no major technological barriers to this scenario – the question is: do we have the collective will to bring it about?

ACKNOWLEDGEMENT

Author would like to acknowledge IST projects SATIN, MoDiS and MAESTRO and ESA project BGAN extension for the material used in this paper.

REFERENCES

[1] A.Franchi, E.Trachtman, L.Christodoulides, J.Sengupta; "Multimedia via Inmarsat"; IEEE Multimedia; Vol. 6, No. 4, pp 15-19, Oct.-Dec. 1999.
[2] http://www.ee.surrey.ac.Uk/Personal/L.Wood/ constellations/ overview.html; as appeared in April 2004
[3] P.S.Aston; "Satellite telephony for field and mobile applications"; IEEE Aerospace Conference Proceedings; Vol. 1, pp 179-190; March 2000
[4] http://www.thuraya.com/; as appeared in April 2004

[5] ETSI TS 101 376-1-3 V1.1.1 (2001-03); "GEO-Mobile Radio Interface Specifications"; Part 1: General specifications; Sub-part 3: General System Description; GMR-1 01.202

[6] A.Franchi, A.Howell and J.Sengupta; "Broadband mobile via satellite: Inmarsat BGAN"; IEE Seminar on Broadband Satellite: The Critical Success Factors - Technology, Services and Markets; pp 23/1 - 23/7; October 2000

[7] http://www.ist-satin.org/; as appeared in April 2004

[8] K.Narenthiran, M.Karaliopoulos, B.G.Evans, et al; "S-UMTS access network for broadcast and multicast service delivery: the SATIN approach"; International Journal of Satellite Communications and Networking, Vol. 22, No. 1, pp 87-111; Jan.-Feb. 2004.

[9] 3GPP TS 22.146 V6.3.0 (2004-01); "Multimedia Broadcast/Multicast Service";Stage 1 (Release 6)

[10] 3GPP TS 22.246 V6.0.0 (2004-01); "Multimedia Broadcast/Multicast Service (MBMS) user services";Stage 1 (Release 6)

[11] 3GPP TS 23.246 V.6.1.0 (2003-12); "Multimedia Broadcast/Multicast Service (MBMS)"; Architecture and functional description
(Release 6)

[12] http://www.ist-modis.org/ as appeared in April 2004

[13] K.Narenthiran, M.Karaliopoulos, R.Tafazolli, B.G.Evans, et al; "S-DMB System Architecture and the MODIS Demo", IST Mobile & Wireless Summit; June 2003 (June); Aveiro, Portugal

[14] http://www.ist-maestro.dyndns.org/ as appeared in April 2004

[15] 3GPP TS 25.401 V6.2.0 (2003-12); "UTRAN overall description" (Release 6)

[16] http://www.mbco.co.jp/english/

[17] A.Hale and D.Ballinger "Military applications for Digital Audio Radio Service (DARS)"; IEEE Aerospace Conference Proceeding; Vol. 3, pp 3-1039 - 3-1050; March 2002.

[18] http://www.connexionbyboeing.com/ as appeared in April 2004

[19] W.H.Jones, M.de La Chapelle; "Connexion by Boeing[SM]-broadband satellite communication system for mobile platforms"; IEEE Military Communications Conference (MILCOM 2001); Vol. 2; pp 755-758; Oct. 2001.

[20] http://www.comnets.rwth-aachen.dc/~o_drive/ index. html as appeared in April 2004

[21] ETSI EN 302 307 V1.1.1 (2004-01); Digital Video Broadcasting (DVB); "Second generation framing structure, channel coding and modulation systems for Broadcasting, Interactive Services, News Gathering and other broadband satellite applications"

[22] WWRF book of vision 2004 (To be published soon)

PART IV

ENCODING, ALGORITHMS AND
PERFORMANCE

Chapter 17

APPLYING NEAR SHANNON-LIMIT CODES TO WIRELESS COMMUNICATIONS

MUSTAFA EROZ, LIN-NAN LEE, FENG-WEN SUN
Hughes Network Systems, 11717 Exploration Lane, Germantown, Maryland 20854, U.S.A.

Abstract: Tremendous progress has been made on channel coding techniques during the last decade, starting from the invention of turbo codes and the rediscovery of the low-density parity check (LDPC) codes. These codes essentially closed the gap between the Shannon capacity limit and practical implementation. Commercial adoption of these new coding techniques, however, depends heavily on the introduction of the technology into communication standards. In this paper, we will briefly review the considerations when turbo codes and LDPC codes were introduced to the 3rd generation wireless standards (3GPP& 3GPP2), and the digital video broadcast for satellite standard (DVB-S2), respectively.   We will then examine a few possible approaches to take advantage of these powerful codes for future wireless channels.

1.   INTRODUCTION

Forward error correction technology remained static for almost a quarter century since the landmark invention of Viterbi decoding for convolutional codes in 1968 [1].   Coding theorists of the time had considered the computation cutoff rate, R_{comp}, instead of the Shannon capacity as the practical limit one can establish reliable communications [2]. This is exactly why all the second generation cellular systems are based on convolutional codes and Viterbi decoding. The discovery of turbo coding in 1993 [3], not only proved that Shannon was correct all along, but it also led us to introduce turbo codes to the third generation (3G) wireless standards in the late 1990s into both 3GPP and 3GPP2 [4]. While the LDPC codes had been discovered by R. Gallager in the early 1960s [5], the success of iterative decoding [6] of turbo codes motivated the rediscovery of LDPC codes in the late 1990s [7], which became a strong alternative to turbo codes in many applications. It has been demonstrated that Shannon limit can be achieved with infinitesimal degradation with sufficiently long LDPC codes [8].   This led us to investigate the feasibility of achieving the Shannon capacity with reasonable implementation complexity using the current technology.

The second generation Digital Video Broadcast via Satellite standard (DVB-S2) is another landmark activity after the 3rd generation wireless standards where the best of coding technology can be applied on a world-wide scale commercially [9].   The original digital video broadcast via satellite standard (DVB-S) was developed in early 1990s [10], based on concatenation of convolutional codes with Reed-Solomon codes [11] and has been wildly successful. The objective of DVB-S2 is to increase the capacity of satellite broadcast channel by at least 30 percent over DVB-S with no greater than 14 mm^2 of silicon area using 13-μ design rules while all link parameters are kept the same. The selected codes based on concatenation of LDPC and BCH codes exceed the original capacity by 35 to 40 percent [12]. As demonstrated by Figure 1, their performance is about 0.7 dB away from the Shannon capacity. While this phenomenal result is achieved for the satellite broadcast channel, which is considerably friendlier than the wireless channels, it is our expectation that the new generation of LDPC based codes can make significant impact on wireless applications. In this chapter, we will discuss essential considerations for designing near Shannon-limit codes in the wireless channel.

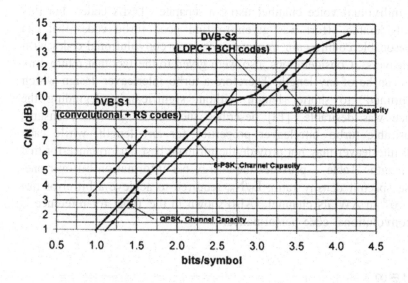

Figure 1. Channel Capacity of DVB-S2 and DVB-S as function of operating C/N

2. DESIGN CONSIDERATIONS FOR 3G TURBO CODES

Instead of Additive White Gaussian Noise (AWGN), the dominant performance limitation in the wireless channel is multipath Rayleigh fading. When applying turbo codes to a fading channel in our earlier work, the power advantage of turbo codes over a conventional convolutional coding scheme in the AWGN channel rapidly diminishes. An example is shown in Figure 2. It was then realized that since all the Code-Division Multiple Access (CDMA) based wireless standards implement very tight power control to counter the near-far problem, and to achieve good multiple access capacity, most of the power advantage of turbo codes in fact can be recovered in a power controlled Rayleigh fading channel, as indicated in Figure 3.

In the 3rd generation wireless development, we need to support voice services with each individual voice channel using a separate CDMA code. For this reason, considerable effort was given to find good turbo codes that can provide reasonable coding gain improvements over convolutional codes with short interleavers. Considerations also were given to the fact that return link data bursts can be of any length. Therefore a turbo interleaver design which can be "pruned" to shorter length without significantly degrading the performance was desirable. [13] describes such a design which ensures certain desirable turbo interleaver properties for *any* length, while each individual interleaver may not provide the ultimate best performance for that particular length. Table 1 provides the energy per information bit over one-sided noise spectral density ratio, E_b/N_o, required to achieve a bit error rate (BER) of 10^{-3} in AWGN channel. Also shown in the Table for reference is the K=9 convolutional code performance.

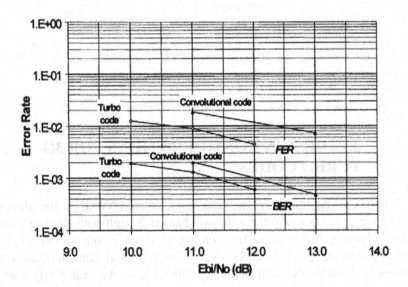

Figure 2. Comparison for bit error rate and frame error rate between rate ½ turbo and convolutional codes in a Rayleigh fading channel without power control at 30 kmph vehicular speed, 3072-bit frame

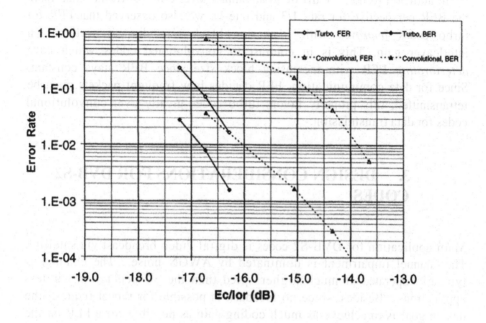

Figure 3. Comparison for bit error rate and frame error rate between rate 1/3 turbo and convolutional codes in a Rayleigh fading channel with power control at 30 kmph vehicular speed, 1280 bit frame, 78,6 kbps, 8dB I_{or}/I_{oc}*Table 1.* E_b/N_0 (dB) required to achieve a bit error rate (BER) of 10^{-3} for 3G turbo codes in AWGN channel as a function of interleaver length in information bits

Table 1

Rate	Convolutional Code	Turbo Codes Interleaver Length			
		512	1024	2048	3072
½	2.2	1.5	1.3	1.1	1.0
1/3	1.9	0.8	0.6	0.4	0.3
¼	1.7	0.6	0.3	0.2	0.1

It can be observed that by doubling the interleaver length, we can increase the coding gain by about 0.2 dB. We should caution, however, that a BER of 10^{-3} is by no means 'reliable" communications in the sense of Shannon theory, but we used it as a practical objective for wireless applications.

In addition to the 1.1 dB or so advantage over convolutional code from the BER perspective for rate 1/3 and rate ¼, we also observed that FER for turbo codes *improves* with increasing block size due to well-known interleaver gain. This is in contrast to convolutional codes which have *deteriorating* FER with increasing block size since BER stays constant. Since for data communication, FER decides how frequent packets must be retransmitted, this property favors turbo codes greatly over convolutional codes for data transmission.

3. DESIGN CONSIDERATIONS FOR DVB-S2 CODES

Main application for DVB-S2 codes is digital video broadcast via satellite. The channel impairment is dominated by AWGN noise. The channel is typically operated at much higher speed than the 3rd generation wireless applications. Besides, since no feedback is possible for repeat request, the design goal is to achieve as much coding gain as possible for a FER on the order of once per hour at threshold. These considerations push us to use as long a code frame as practical. As the new coding schemes based on iterative decoding are much more powerful than the original DVB-S codes, an important consideration is to achieve higher bandwidth efficiency and higher throughput for typical satellite broadcast links. This is a different tradeoff from the 3G coding consideration, as they typically need to be achieved with higher rate codes. This led us to consider LDPC codes of length 64,800 bits concatenated with very high rate BCH codes. The LDPC codes are selected based on the smallest distance from Shannon capacity among seven candidate proposals. It is about 0.3 dB better than the next best candidate based on turbo codes concatenated with Reed-Solomon codes. Figure 4 shows the performance of these concatenated LDPC and BCH codes in the AWGN channel, when combined with QPSK, 8-PSK, 16-APSK, and 32-APSK.

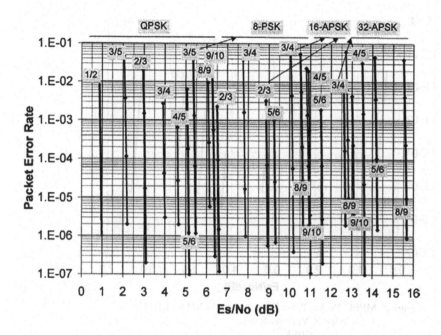

Figure 4 MPEG Packet Error Rate for the DVB-S2 codes with a
number of modulation format

Recognizing that these codes may be too long for some of the non-broadcast applications such as interactive and satellite news gathering, a set of LDPC codes of length 16,200 bits were also designed and selected. These codes generally require about 0.2-0.25 dB more power than the full-length LDPC codes of the same rate, but considerably simpler to implement. Its performance is about the same as, or slightly better than the best turbo code candidate proposed earlier. Figure 5 shows the performance of these codes.

We also designed an LDPC code with block length reduced to 8,100 bits. As shown in Figure 6, for rate ½, the code performs quite similar to the corresponding turbo code with similar interleaver length (3072 information bits, or 6144 coded bits) at BER of 10^{-3} or 10^{-4}. We can conclude that LDPC codes are apparently more attractive for longer block length.

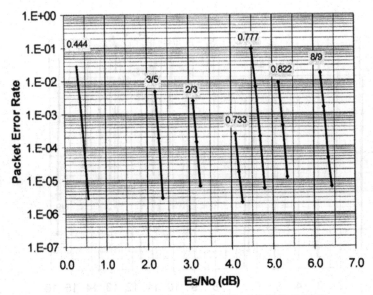

Figure 5 MPEG Packet Error Rate for N = 16200 bit LDPC
Codes of Various Rates,
QPSK.

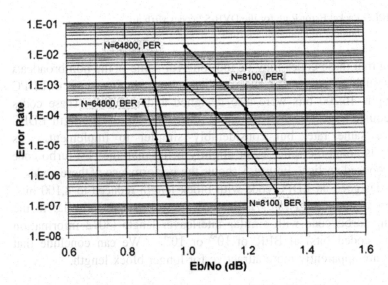

Figure 6 BER and MPEG Packet Error Rate (PER) of rate ½
LDPC codes, of length 8100 bits and the DVB-S2 code of length
64800 bits.

4. ENHANCEMENTS OF 3G WIRELESS STANDARDS IN THE FORWARD LINK

With the development of cdma2000 1xEVDO, cdma2000 1xEVDV and HSDPA enhancements by both 3GPP2 and 3GPP, the maximum speed of the forward link for the 3rd generation wireless channel is greatly increased. When operating at maximum speed, a high-speed TDM channel with little processing gain is used to send high-speed data to the user terminals, very much like what we have proposed in [14]. These schemes have also gone a step further by using m-ary QAM modulation for the highest data rate. Transmission of high-speed data in such high-order modulation with little or no spectral spreading, obviously, is backed up by rate adaptation. Though quite a departure from earlier experiences in the 1st and 2nd generation systems, these schemes can be successful in the wireless environment because

- Maximum speed is achieved only when close to the base station, where signal-to-noise ratio is high, and other cell interference is low
- When operating at near maximum speed, delay spread spreads the multipath energy to more than one symbol away, classical equalizers instead of rake receivers can also be effective
- Channel characteristics typically do not change significantly within an interleaver frame or code block

Power control in such a high-speed forward link channel is generally not very effective, and rate adaptation is used to ensure that all the terminals can in fact receive information at data rate supportable by the link. With these characteristics, it is not difficult to imagine that the high-speed forward link channel in the wireless environment is not significantly different from the digital satellite broadcast channel for which the DVB-S2 codes are designed for. Of particular interest may be with those coding/modulation combinations delivering more than 2 bit/symbol. With these codes delivering performance near the Shannon limit, they do appear quite attractive.

5. ENHANCEMENT OF THE 3G WIRELESS RETURN LINK

Data traffic tends to be asymmetrical. Traffic burst tends to be in short bursts in return links. Long LDPC codes do not present themselves as attractive options in the return link. Power efficiency, however, translates directly to battery talk-time, and is therefore very important. For a CDMA system, we have the freedom of using lower rate codes, since the signal is spread into a wider bandwidth anyway. We can also tradeoff spreading factor with code rate for the same transmit duration. Comparing cases with 512 information bit cases, about 1 dB coding gain can be had when the code rate is reduced from rate ½ to rate ¼ from BER perspective. By lowering the code rate for short bursts, we can easily compensate the inadequate performance in the short burst cases. Since the bursts are short, they do not require very much in-route capacity, even when it is coded with very low rate codes. By doing this, we create a new concept: rate adaptation to equalize frame error rate (FER). In the AWGN channel, [4] shows the E_b/N_o required to achieve 10^{-2} FER of 3G turbo codes as summarized in Table 2 below.

Table 2 E_b/N_o (dB) required by 3G turbo codes to achieve 10^{-2} FER

Rate	Frame length	
	512	3072
½	1.6	1.1
¼	0.7	-0.1

As CDMA system capacity is directly related to the operating E_b/N_o, if we can lower the operating point of the system by equalizing the frame error rate for bursts of all length, the overall capacity can be improved. From Table 2, it is clear that by simply adapting rates between ½, 1/3, and ¼, for different length of bursts, at least 0.5 dB or about 12 percent of additional system capacity can be achieved. By equalizing the FER, we also reduce the number of retransmission of shorter bursts, and thereby improve the system performance from user perspective. This simple exercise points to the direction for further improving the reverse link performance by designing good codes with rate lower than ¼ in the future.

6.   EXTENDING AWGN RESULTS TO MULTI-INPUT, MULTI-OUTPUT CHANNELS

We have argued in the previous sections that for power controlled CDMA channel, or for high-speed data channels with little or no spectrum spreading, codes that perform well in the AWGN channel continues to be the most attractive candidates in wireless channels. It can be seen that this is also true for receive and transmit diversity. By applying LDPC codes on the forward link, and turbo codes on the return link the most desirable performance can be achieved with transmit and/or receive diversity. Recent wireless research has been focusing on space-time multi-input and multi-output (MIMO) transmission, particular for the forward link, where high capacity is the most desirable. The original space-time designs [15, 16] were based on short constraint length convolutional codes. While these codes achieve maximum diversity given the antenna configuration, the error correcting capability of these convolutional codes is no match for the state-of-the-art turbo codes or LDPC codes. It is natural to investigate the possibility to design space-time codes based on turbo or LDPC codes. We find it more difficult to design turbo-based space time codes, since the complexity of the constituent code increases exponentially with number of transmit antennas. Also, it is very difficult to configure the random interleaver that may exploit the n-transmit, m-receive antenna configuration. For LDPC codes, however, the complexity is determined by the length of the code. Further, it is possible to design the parity check matrices_of the LDPC codes to avoid any singularity that could be detrimental to space-time code performance.

Another alternative is to separate the function of space-time codes into two parts. The first part tries to achieve maximum diversity given the antenna configuration, and the second part tries to provide the error correction capability. Thus, diversity and error correcting performance can be optimized independently. In this case, we believe both turbo and LDPC codes can be very effective candidates for the second part due to their near-Shannon limit performance. But, because the high number of iteration afforded by a LDPC decoder compared to a turbo decoder of the same complexity, LDPC decoder is easier to recover from a burst of very poorly received bits compared to turbo codes.

7. SUMMARY

We believe that a coding technique which achieves near-Shannon limit performance in the AWGN channel will still perform the best in the wireless channel. In the case of return links, turbo codes offer a clear advantage. In the case of forward link, with or without MIMO, LDPC codes seem to offer greater advantage either by straightforward concatenation with a space-time transformation which achieves full-diversity given the antenna configuration, or with direct construction by taking account of the channel configuration

REFERENCES

[1] A.J. Viterbi, "Error Bounds for Convolutional Codes and An Asymptotically Optimal Decoding Algorithm," IEEE Transactions on Information Theory, IT-13, pp. 260-269, April 1967.

[2] L. Lee, "On Optimal Soft-Decision Demodulation," IEEE Transactions on Information Theory, Vol. IT-22, No.4, July 1976, pp.437-444.

[3] C. Berrou, A. Galvieux, and P. Thitimajshima, "Near Shannon Limit Error Correcting Coding and Decoding: Turbo Codes," Proceedings of ICC (Geneva, Switzerland), May 1993.

[4] L. Lee, R. Hammons, F. Sun, and M. Eroz, "Application and Standardization of Turbo Codes in Third-Generation High-Speed Wireless Data Services," IEEE Transactions on Vehicular Technology, Vol. 49, No.6, November, 2000, pp-2198-2207.

[5] R. G. Gallager, "Low density parity check codes", IRE Trans. Info. Theory, 1962, IT-8, pp. 21-28

[6] L. Lee, "Concatenated Coding System Employing a Unit-Memory Convolutional Code and a Byte Oriented Decoding Algorithm," IEEE Transactions on Communications, Vol. COM-25, No. 10, October 1977.

[7] D. J. MacKay and R. M. Neal, "Good codes based on very sparse matrices", 5th IMA Conf. 1995, pp. 100-111

[8] S. Chung, G.D. Forney, T. Richardson and R. Urbanke "On the Design of Low Density Parity Check Codes within 0.0045 dB of the Shannon Limit", IEEE Comm. Letters, vol. 5, no. 2, Feb. 2001, pp. 58-60.

[9] M. Eroz, F. Sun, and L. Lee, "DVB-S2 Low-Density Parity Check Codes with near Shannon Limit Performance," to appear in International Journal of Satellite Communications, 2004.

[10] U. Reimers (ed.), "Digital Video Broadcasting – The DVB Family of Standards for Digital Television," 2nd ed., 2004, Springer Publishers, New York, ISBN 3-540-43545-X.

[11] J. Odenwalder, "Optimal Decoding on Convolutional Codes," Ph.D. Dissertation, School of Engineering and Applied Sciences, University of California, Los Angles, 1970.

[12] ETSI EN 302 307 V1.1.1 (2004-01) "Digital Video Broadcast (DVB) Second Generation Framing Structure, Channel Coding and Modulation Systems for Broadcasting, Interactive Services, News Gathering and Other Broadband Satellite Applications."

[13] M. Eroz and A. R. Hammons Jr., "On the design of prunable interleavers for turbo codes", *in Proc. VTC'99, May 16-19, 1999, Houston, TX*

[14] L. Lee, K. Karimullah, F. Sun, and M. Eroz, "Third Generation Wireless Technologies – Expectation and Realities," Proceedings, PIMRC'98, September 8-11, 1998, Boston, MA.

[15] V. Tarokh, N. Seshadri and A.R. Calderbank, "Space-time codes for high data rate wireless communication: performance criterion and code construction", IEEE Trans. Info Theory, vol. 44, no.2, pp. 744-764, March 1998.

[16] A.R. Hammons and H. El Gamal, "On the theory of space-time codes for PSK modulation", IEEE Trans. on Info Theory, vol. 46, no.2, pp. 524-542, March 2000.

Chapter 18

TIME-DIFFUSION CONCEPTS AND PROTOCOL FOR SENSOR NETWORKS

WEILIAN SU[1], IAN F. AKYILDIZ[2]

[1]*Departmenl of Electrical and Computer Engineering, Naval Postgraduate School, Monterey, CA 93943, USA Email: weilian@ece.gatech.edu;* [2]*Broadband and Wireless Networking Laboratory, School of Electrical and Computer Engineering, Georgia Institute of Technology, Atlanta, GA 30332, USA Tel: (404) 894-5141 Fax: (404) 894-7883 Email: ian@ee.gatech.edu*

Abstract: In the near future, small intelligent devices will be deployed in homes, plantations, oceans, rivers, streets, and highways to monitor the environment. These devices require time synchronization, so voice and video data from different sensor nodes can be fused and displayed in a meaningful way at the sink. The *Time-Diffusion Synchronization Protocol* (TDP) is proposed as a network-wide time synchronization protocol. It allows the sensor network to reach an equilibrium time and maintains a small time deviation tolerance from the equilibrium time. In addition, simulations are performed to validate the effectiveness of TDP in synchronizing the time throughout the network and balancing the energy consumed by the sensor nodes.

1.   INTRODUCTION

In the near future, small intelligent devices will be deployed in homes, plantations, oceans, rivers, streets, and highways to monitor the environment[1]. Events such as target tracking, speed estimating, and ocean current monitoring require the knowledge of time between sensor nodes that detect the events. In addition, sensor nodes may have to time-stamp data packets for security reasons. With time synchronization, voice and video data from different sensor nodes can be fused and displayed in a meaningful way at the sink.

There are few synchronization protocols proposed for the sensor networks. These protocols are *post-facto synchronization*[2], *Reference-*

Broadcast Synchronization (RBS)[3], *Network Time Protocol* (NTP)[4], and *Time-Sync protocol for Sensor Networks* (TPSN)[5]. Most of these proposed protocols provide local synchronization while NTP may be used for network-wide synchronization. Although NTP may be useful to discipline all the sensor nodes, the sensor nodes may be off when power management and topology maintenance protocols, e.g. LEACH[6], are employed. In addition, disciplining all the sensor nodes in the sensor field may be a problem due to interferences from the environment and large variation of delays between different parts of the sensor field. The interferences can temporarily disjoint the sensor field into multiple smaller fields causing undisciplined clocks among these smaller fields.

To provide network-wide time synchronization, the time differences among the sensor nodes must be minimized before protocols requiring time-stamps, e.g., security applications, flow control protocols, target tracking, voice fusion, video fusion, and environmental data fusion, are realizable. In addition, the time synchronization protocol must be robust to node failures as well as energy consumption in the network. Also, node mobility must be taken into account.

As a result, we propose the *Time-Diffusion Synchronization Protocol* (TDP). The motivations for network-wide time synchronization are as follows:

- Enable applications to coordinate sensor nodes, e.g., target tracking, data fusion, and decision fusion.
- Enable users to perceive events in the same time frame, e.g., multiple fire outbreaks at different locations of the sensor field.
- Enable protocols that require time-stamps, e.g., security, flow control, and medium access protocols.

The TDP is used to maintain the time throughout the network within a certain tolerance. The tolerance level can be adjusted based on the application of the sensor networks.

The design issues and TDP are described in Section 2. Afterwards, the simulation results are discussed in Section 3. Lastly, the paper is concluded in Section 4.

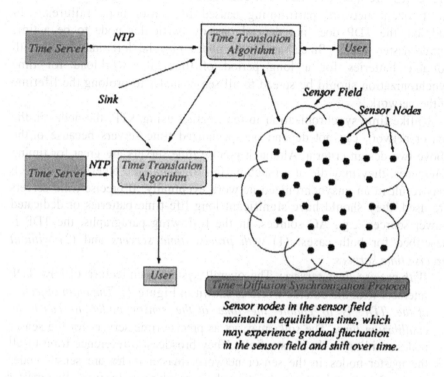

Figure 1 System architecture.

2. DESIGN ISSUES AND TIME-DIFFUSION SYNCHRONIZATION PROTOCOL

Small and low-end sensor nodes may exhibit device behaviors that may be much worse than the large systems such as *personal computers (PCs)*. As a result, the time synchronization in these nodes presents a new challenging problem. Some of the factors influencing time synchronization in large systems also apply to sensor networks[7]; they are *temperature, phase noise, frequency noise, asymmetric delays,* and *clock glitches.*

Besides dealing with these factors, a time synchronization protocol for sensor networks should be *automatically self-configured* and be *sensitive to energy requirement.* These are the two design criteria that the TDP is engineered around. The TDP self-configures and self-organizes to address the frequent network partitioning caused by sensor node failures. In addition, the TDP does not depend on any particular node to be a time server/master node. Since the sensor nodes may be left unattended with portable batteries for a long period of time, the workload for time synchronization should be spread to all sensor nodes to prolong the lifetime of the network.

Unlike time synchronization in the Internet using NTP, the nodes in the sensor network can not depend on specialized time servers because of the above two design criteria. Although precise time servers are great for timing references, they may die at a much earlier time than the other nodes, which may result in an unsynchronized network eventually. If precise time servers are used, they should have significant long life-time batteries or dedicated power sources, e.g., AC sources. In the following paragraphs, the TDP is described for both cases: (1) *with precise time servers* and (2) *without precise time servers.*

1. *With precise time servers:* The overall system architecture of how TDP interacts with the outside world is shown in Figure 1. *The main objective of the TDP is to enable the time of the sensor nodes to reach an equilibrium time.* The *sinks* may act as precise time servers for the sensor nodes residing in the sensor field. They broadcast a reference time to all the master nodes in the sensor network; master nodes are sensor nodes randomly elected to synchronize their neighbors. In turn, the master nodes use the received reference time to synchronize their neighbor nodes by using the TDP. In essence, the equilibrium time that the sensor network reaches is the reference time broadcast by the sinks.

2. *Without precise time servers:* Although the TDP can be used with precise time servers, it is more important to discuss about the autonomous nature of TDP since the line-of-sight or connection to all master nodes from the sinks may not be possible. Also, the sensor network may be deployed in areas that may not be accessed by the sinks for a long period of time, e.g., caves and ocean floor. Consequently, the sinks may not be used as time servers; fortunately, the autonomous nature of the TDP allows the sensor network to reach an equilibrium time that is independent from the time used by the Internet, e.g., *Universal Coordinated Time (UTC).*

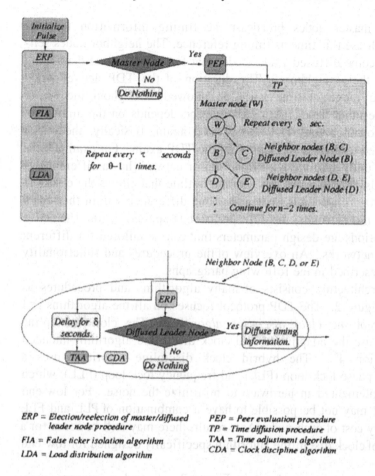

ERP = Election/reelection of master/diffused   PEP = Peer evaluation procedure
      leader node procedure   TP = Time diffusion procedure
FIA = False ticker isolation algorithm   TAA = Time adjustment algorithm
LDA = Load distribution algorithm   CDA = Clock discipline algorithm

Figure 2. TDP architecture.

Since the time in the sensor network reaches an equilibrium value, it still may drift over time and has fluctuation throughout the sensor network. *From the perspective of the outside world, the sensor network is like a multi-dimensional clock, where the time varies in space, i.e., sensor field, and time.* Although the time variation throughout the sensor network may be very small, it is necessary to translate the time in the sensor network to a common time, e.g., UTC, used by the users. The sinks as shown in Figure 1 take care of such translation with the *Time Translation Algorithm* and serve as interfaces to the sensor network.

The time schedule for applying TDP consists of both active and inactive periods. During the active period, the master nodes are reelected at every τ seconds, which is a design parameter that depends on the types of sensor networks. The master nodes broadcast the timing information to their neighbors, which use this time as timing reference. The neighbor nodes self-determine to become diffused leader nodes that further broadcast the timing information to their neighbors. The duration of the TDP active period depends on the range of time variation allowed throughout the sensor network. On the other hand, the inactive period depends on the amount of clock drifts allowed before TDP is activated again. Basically, the sensor nodes are allowed to drift before applying the TDP again. For example, the TDP is applied until the time within the sensor network has a difference of 100 milliseconds. It is inactive for a period of time that allows the clocks to drift apart by 50 milliseconds. When the time difference within the sensor network is 150 milliseconds, the TDP protocol is applied again. The active and inactive periods are design parameters that can be tailored for different types of sensor networks. An overview of the procedures and functionality of the TDP is described in the following paragraphs.

The TDP architecture consists of many algorithms and procedures as illustrated in Figure 2. The TDP protocol focuses on all the algorithms and procedures except the *clock discipline algorithm.* The clock discipline algorithm may use the adaptive hybrid clock discipline algorithm intended for NTP Version 4^8. The hybrid clock discipline algorithm uses a combination of phase lock loop (PLL) and frequency lock loop (FLL), which are usually implemented in hardware to minimize the noise. For low-end sensor nodes, it may not be possible to have a combination of PLL and FLL due to monetary cost of each node. As a result, there may still be room for a different type of clock discipline algorithm specifically designed for low-end sensor nodes.

The algorithms and procedures in Figure 2 are used to autonomously synchronize the nodes, remove the false tickers (clocks that deviate from their neighbors), and balance the load required for time synchronization among the sensor nodes. Initially, the sensor nodes may receive an *Initialize pulse* from the sink either through direct broadcast or multi-hop flooding. Then they self-determine to become master nodes with the *election/reelection of master/diffused leader node procedure (ERP)*, which consists of the *false ticker isolation algorithm (FIA)* and *load distribution algorithm (LDA)* as shown in Figure 2.

The LDA distributes the energy consumption for diffusing timing information messages to all sensor nodes in the network. It achieves them by reelecting master and diffused leader nodes at every τ and δ seconds, respectively. The value δ is the amount of time between each round of timing information message diffusion during the τ period. During the reelection, the nodes randomly choose a value λ that is between 0 and 1.

The value λ is then shifted by the value $(1- \zeta)$, where ζ is the ratio of current energy level over the maximum allowed energy level, and calculated as

$$\lambda = \lambda - (1 - \zeta) \tag{1}$$

If the value λ is greater than the threshold φ, then the node is either a master or diffused leader node depending if the master or diffused leader node is being reelected. The threshold φ determines the number of sensor nodes participating as a master or diffused leader node. For example, if φ is set equal to 0.7, it means on the average that 30 percent of the deployed sensor nodes is a master node or diffused leader node. As a result, $\rho = 1 - \varphi$ represents the fraction of deployed sensor nodes that is a master or diffused leader node. For this case, ρ is set equal to 0.3.

Since the shifting of the randomly selected value λ is based on the current energy level of the sensor node, ρ decreases if the threshold φ is not adjusted appropriately. As a result, the threshold φ stored in all sensor nodes is adjusted at every τ seconds according to

$$\varphi = \varphi - \varepsilon \tag{2}$$

where ε is the amount that needs to be adjusted, which is based on μ (energy consumed per round of timing information message diffusion). The value μ can be approximated by

$$\mu \approx \frac{\text{Amount of energy consumed during } \tau \text{ seconds}}{\lceil \tau / \delta \rceil - 1} \tag{3}$$

where τ is the master node reelection period, and δ is the time between each round of timing information message diffusion.

As a result, the value ε is calculated as

$$\varepsilon = \rho - \sum_{m-1}^{i} \Phi_{i,m} \rho^{m-1} (1 - \rho)^{i-m} (\rho - (m-1)\eta) \tag{4}$$

where ρ is the fraction of sensor nodes that can become a master or diffused leader node; i is the number of rounds within a τ period, which is approximated by $\lceil \tau / \delta \rceil - 1$; η is the ratio of μ (Eq. 3) over the maximum energy level; and the coefficient $\Phi_{i,m}$ is calculated as

$$\Phi_{i,m} = \begin{cases} 1 & , \text{for } m = 1 \\ \displaystyle\sum_{j=1}^{i-1} 1 & , \text{for } m = 2 \quad (5) \\ \left(\displaystyle\sum_{v_{m-1}=1}^{i-(m-1)} \left(\displaystyle\sum_{v_{m-2}=1}^{i-(m-2)-v_{m-1}} \cdots \left(\displaystyle\sum_{v_1=1}^{\binom{i-1-\sum_{k=1}^{m-2} v_{m-k}}{}} 1) \right) \cdots \right) \right) & , \text{for } m \geq 3 \end{cases}$$

with $i \geq m$ and $m-1$ levels of summation for $m \geq 3$, e.g., $\left(\sum\left(\sum\bullet\right)\right)$ and $\left(\sum\left(\sum\left(\sum\bullet\right)\right)\right)$ are 2 and 3 levels, respectively.

At the end of procedure *ERP* as shown in Figure 2, the elected master nodes start the *peer evaluation procedure (PEP)* while others do nothing. The procedure *PEP* helps to remove false tickers from becoming a master node or a diffused leader node. It uses the 2-sample Allan variance $\sigma^2(\iota)$ [4,8] to determine if the local clocks are deviated from each other. The 2-sample Allan variance $\sigma^2(\iota)$ is calculated as follows:

$$\sigma^2(\iota) = \frac{1}{2\iota^2(N-2)} \sum_{g=1}^{N-2} (x_{g+2} - 2x_{g+1} + x_g)^2 \tag{6}$$

where ι is the time difference between two time deviation measurements; N is the total number of time deviation measurements, and x is the measurement value.

After procedure *PEP,* the elected master nodes (denoted by W in Figure 2) start the *time diffusion procedure (TP),* where they diffuse the timing information messages at every δ seconds (round interval) for a duration of τ seconds. Each neighbor node (e.g., node B or C in Figure 2) receiving these timing information messages self-determines to become a diffused leader node using the procedure *ERP.* Furthermore, all neighbor nodes adjust their local clocks using *time adjustment algorithm (TAA)* and *clock discipline algorithm* (CDA) after waiting for δ seconds as shown in Figure 2.

The elected diffused leader nodes (e.g., node *B*) will further diffuse the timing information messages to their neighboring nodes (e.g., nodes *D* and *E*) within their broadcast range. Note that these timing information messages are diffused by each elected diffused leader node for *n* hops from the master nodes, where each hop represents one level from the master nodes (e.g., nodes *B* and *C* are at Level-*1* while nodes *D* and *E* are at Level-*2*). This diffusion process enables all nodes to be autonomously synchronized. In addition, the master nodes are re-elected at every τ seconds using the procedure *ERP*, which is repeated for $\theta - 1$ times, where $\theta\tau$ is equal to the length of the TDP active period.

3.   PERFORMANCE EVALUATION BY SIMULATION

The performance of the TDP is evaluated with an event driven simulation. Two hundred sensor nodes are deployed randomly in an 80 meters by 80 meters sensor field. Each of the sensor nodes can receive and transmit messages to its neighbors by executing the TDP independently, i.e., each sensor node is emulating a physical sensor node where it has its own memory. In addition, it keeps track of its own local time with a randomly selected drift rate that is between ± 100 ppm. Since each node keeps track of its local time, simulations with large number of nodes, e.g., 1000, 2000, and 3000, may become difficult. It is because the simulation has to create an event for every clock tick. As a result, only 200 sensor nodes are deployed with the targeted precision of 10^{-1} seconds order. To show that TDP is able to reach its equilibrium time and maintain a small variation of the deviated time throughout the network, the local time of each sensor node is initially shifted by a random amount ranging from 10 seconds to 60 seconds from the ideal time. Since the local times of the neighbor nodes are quite different, this setup also shows how TDP recovers from network partitioning. In essence, this setup represents the worst case scenario in synchronizing time, where each node may be drifted far apart from each other.

The performance of TDP is evaluated and compared to *Time-Sync Protocol for Sensor Networks* (TPSN)[5] to show its novelties in Section 3.1.

Chapter 18

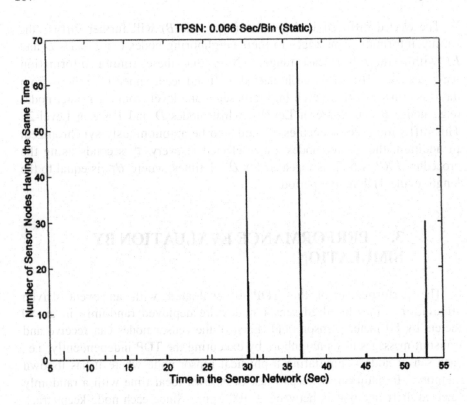

Figure 3 TPSN: histogram of time distributed in the network (static nodes).

3.1 TDP vs. TPSN

The performance of TDP is compared with TPSN in a network-wide scenario to show how the diffusion process helps to synchronize the time in the network.   For TPSN, there are three sinks trying to synchronize the network.   After the nodes are synchronized, the histogram of the sensor nodes' time is calculated and shown in Figure 3. There are three large islands of time occurring approximately at 30 sec, 37 sec, and 54 sec. These islands of time are known to occur[5] when three sinks are used to synchronize the network. These islands of time may cause problems when the users want all sensor nodes to perform a task at a specific time. Although most of the

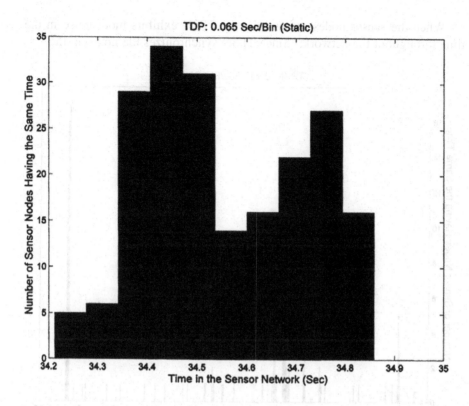

Figure 4 TDP: histogram of time distributed in the network (static nodes).

sensor nodes are synchronized to either one of the three sinks, there are still some nodes that remain unsynchronized. From example, some of the sensor nodes have time values that are within the range of 5 sec and 27 sec. This anomaly may be due to (1) the broadcast radius not being large enough and (2) the timing offset of synchronization messages between two levels in the hierarchy.

Under the same simulation scenario, the TDP is applied. Since the TDP does not depend on specific sensor nodes to be master nodes, it enables the network time to reach an equilibrium value by diffusion process. As shown in Figure 4, the equilibrium time is around 34 sec. The time variation throughout the network is around 0.6 sec. This variation may be much tighter when the master nodes are synchronized to a time server.

When the sensor nodes are mobile, the TPSN exhibits more noise in the time throughout the network. Since TPSN synchronizes the nodes in the

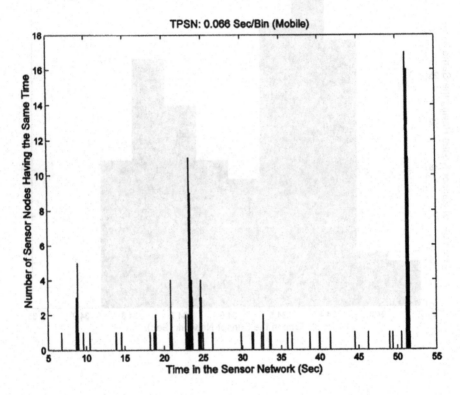

Figure 5 TPSN: histogram of time distributed in the network (mobile nodes).

network hierarchically, the node movement breaks the hierarchy causing nodes to be unsynchronized. As shown in Figure 5, there are still three islands of time but more nodes are becoming unsynchronized due to the movements. As for TDP, the movement does not affect the diffusion process. The time throughout the network still reaches an equilibrium value.

4.   CONCLUSION

The constraint of requiring the nodes to maintain a similar time among the neighbors and throughout the network at conditions where outside timing sources, e.g., high power stations used to discipline the local time of the nodes in the network, may not be available due to distance and location, e.g., inside a cave or under water. With this constraint in mind, we develop the time-diffusion synchronization protocol (TDP) that allows the nodes in the sensor field to reach an equilibrium time with a small tolerance from each other. We have studied the TDP for both static and mobile sensor nodes. In both scenarios, the TDP enables the time in the network to converge to the targeted tolerance, and it outperforms TPSN.

REFERENCES

1. I. F. Akyildiz, W. Su, Y. Sankarasubramaniam, and E. Cayirci, Wireless Sensor Networks: A Survey, *Computer Networks (Elsevier) Journal,* 38(4), 393-422 (2002).
2. J. Elson and D. Estrin, Time Synchronization for wireless sensor networks, *In Proceedings of the 15th International Parallel and Distributed Processing Symposium (IPDPS-01),* IEEE Computer Society, April 2001.
3. J. Elson, L. Girod, and D. Estrin, Fine-Grained Network Time Synchronization using Reference Broadcasts, *Proceedings of the Fifth Symposium on Operating Systems Design and Implementation (OSDI 2002),* Boston, MA, December 2002.
4. D. L. Mills, Adaptive Hybrid Clock Discipline Algorithm for the Network Time Protocol, *IEEE/ACM Trans. on Networking,* 6(5), 505-514 (1998).
5. S. Ganeriwal, R. Kumar, and M. B. Srivastava, Timing-Sync Protocol for Sensor Networks, *ACM SenSys 2003,* Los Angeles, CA, November 2003.
6. W. R. Heinzelman, A. Chandrakasan, and H. Balakrishnan, Energy-Efficient Communication Protocol for Wireless Microsensor Networks, *IEEE Proceedings of the Hawaii International Conference on System Sciences,* 1-10 (2000).
7. J. Levine, Time Synchronization Over the Internet Using an Adaptive Frequency-Locked Loop, IEEE Transaction on Ultrasonics, Ferroelectrics, and Frequency Control, 46(4), 888-896 (1999).
8. D. Allan, Time and Frequency (Time-Domain) Characterization, Estimation, and Prediction of Precision Clocks and Oscillators, *IEEE Trans. on Ultrasonics, Ferroelectrics, and Frequency Control,* 34(6), 647-654 (1987).

Chapter 19

INTERFERENCE, INFORMATION AND PERFORMANCE IN LINEAR MATRIX MODULATION

Olav Tirkkonen and Mikko Kokkonen

Nokia Research Center, Box 407, FIN-00045 NOKIA GROUP, Finland

Abstract: The choice of basis for linear matrix modulation (linear space-time code with rotated and linearly combined constellation) is considered. Unitarily invariant polynomials of square matrices are discussed, the full spectrum of invariants interpolating between the well-known trace and determinant. These give the full spectrum of space–time code design criteria. The diagonal dominance (expansion around the trace) of these invariants is considered. Using this, it is shown that minimizing the self-interference, or equivalently, maximizing the second order expansion coefficient of the mutual information around SNR=0, is required when maximizing the mutual information and/or optimizing performance at *any* SNR. As an example, symbol rate 3 schemes for 4 transmit antennas are considered.

1. INTRODUCTION

The first approaches to reach the channel capacities promised by Multiple-Input-Multiple-Output (MIMO) channels[1, 2] were based on transmitting a vector of complex modulation symbols from the multiple antennas[1-4] with possibly concatenated layered channel coding (D-BLAST, V-BLAST)[1, 4]. More recently, inspired by the orthogonal space-time block codes of Tarokh et al.[5], high rate non-orthogonal matrix transmissions, with and without concatenated channel codes have been discussed[6-13]. All of these consider linear transmissions using a set of basis (dispersion) matrices. Compared to vector transmissions, explicit transmit diversity has been added by extending the transmission of each symbol over multiple channel uses.

These works show that high SNR optimization of linear space–time codes is conceptually (but not algorithmically) rather simple. High-SNR performance is characterized by the degree of transmit diversity provided by the transmission. With rates higher than maximum orthogonal design rates, diversity properties are compromised when the modulation constellation transmitted from a given antenna during a given channel use is constrained to be a well-known equidistant and equiprobable PSK/QAM constellation. Transmit diversity and high-SNR performance are optimized by changing the constellation by rotating some symbols, and by mixing symbols that are transmitted on different basis matrices. Here "rotating" indicates rotating in the two-dimensional complex plane, and "mixing" indicates constructing orthogonal linear combinations. The combined action of these can be described in terms of a group of rotations in a higher dimensional space. Good rotations can be sought among discrete [8, 12] or continuous[9-11,14] rotation groups. Such mixing and rotating of symbols affects directly the phase and amplitude of the channel symbols that are transmitted from a given antenna during a given channel use, so that the channel symbol constellation that modulates the carrier changes from antenna to antenna and channel use to channel use, and typically is not taken from a well-known equidistant and equiprobable modulation constellation. Thus it is more appropriate to talk about linear space–time or matrix modulation than of linear space–time codes, when discussing high rate linear space–time transmissions.

Before performance can be optimized, however, a basis has to be chosen. When the symbol rate is less than the number of transmit antennas, this choice is non-trivial, and many non-equivalent alternatives exist. Most basis choices in the literature are ad hoc, or leave the mathematical structure of good bases unclear. Hassibi and Hochwald[7], performed a gradient search in the space of all possible basis matrices, with the target to maximize the mutual information. Heath & al.[6, 11] used frame theoretical bounding tools. Both approaches leave the structure of the resulting codes is rather non-transparent. El Gamal and Damen[12] chose the basis by the heuristic principle of "threading", implying that on each thread, a single-antenna code can be used, and the only-space–time coding aspect is to patch the threads together.

In this paper, we continue the systematic analysis of basis choice, started in[10, 13]. A closed form equation was found, leading to maximization of the second order coefficient of mutual information in an expansion in SNR[10], and a systematic group theoretical method to find solutions of these equations was presented[13]. When investigating solutions of these equations[14] it was found that second order mutual information predicts low-SNR performance very well. Also, it was observed that after optimizing at high SNR using constellation rotations, a modulation with higher second order information outperforms one with lower. Here we shall prove that optimizing second order mutual information affects both mutual information and performance at *all values* of SNR. The reason for this is the diagonal dominance of the positive definite Hermitian matrices characterizing information and performance.

As an example, symbol rate 3 matrix modulation schemes for 4 transmit antennas and block length 4 will be considered.

2. LINEAR MATRIX MODULATION

A matrix modulation with N_t Tx antennas extending over T channel uses is a $T \times N_t$ matrix $\mathbf{X}$. With N_r receive antennas and no multipath propagation, the signal model is

$$\mathbf{Y} = \mathbf{X}\,\mathbf{H} + \text{noise} . \tag{1}$$

Here $\mathbf{Y}$ is the $T \times N_r$ matrix of received signals, and $\mathbf{H}$ is the $N_t \times N_r$ channel matrix. In concrete calculations, the elements of $\mathbf{H}$ are assumed i.i.d., with zero mean complex Gaussian (Rayleigh) distribution. The additive noise is Gaussian. Perfect channel state information (CSI) is assumed at the receiver, and no CSI at the transmitter.

The code matrix $\mathbf{X}$ is a linear function of the K complex modulation symbols x_k, $k = 1 \ldots K$ to be transmitted. The symbol rate is the number of complex symbols transmitted per channel use, K/T. Analytical expressions become simpler when expressed in terms of real symbols c_k, $k = 1, \ldots, 2K$. For QPSK and conventional QAM modulations, the real and imaginary parts may indeed be considered as independent symbols. Due to linearity, the code matrix may be expanded as

$$\mathbf{X} = \sum_{k=1}^{2K} c_k\,\mathbf{B}^{(k)} . \tag{2}$$

The matrices $\mathbf{B}^{(k)}$ are constant matrices with complex entries, and they constitute the basis for the matrix modulation. A priori there are altogether $2KN_tT$ complex degrees of freedom when choosing a basis.

Distance matrix and Chernoff bound. Performance of a space-time code is determined in terms of unitarily invariant polynomials (see Section 3) of elements of the distance matrix $\mathbf{M}^{(ce)} = \mathbf{D}^{(ce)\dagger}\mathbf{D}^{(ce)}$, where $\mathbf{D}^{(ce)}$ is the codeword difference matrix. Here these polynomials shall be shortly referred to as unitary invariants. The best known are the rank[15], the determinant[16], and the trace[17, 18]. Similarly, mutual information properties of linear matrix modulations are determined by unitary invariants of the squared code matrix $\mathbf{X}^\dagger\mathbf{X}$. Due to linearity, properties of the codeword difference matrices and properties of the matrix modulation itself are directly related, $\mathbf{D}^{(ce)}(\boldsymbol{\Delta}) = \mathbf{X}\left(\mathbf{c}^{(c)}\right) - \mathbf{X}\left(\mathbf{c}^{(e)}\right) = \mathbf{X}(\boldsymbol{\Delta})$, where $\Delta_k = c_k^{(c)} - c_k^{(e)}$ are the symbol differences between the transmitted code word $\mathbf{c}^{(c)}$ and the possibly erroneous detected code word $\mathbf{c}^{(e)}$.

The Chernoff upper bound of the union bound of pairwise error probabilities is given by[16]

$$P_{\text{UB}} = N_e\, \mathcal{E}\left\{\left(\mathbf{I}_{N_t} + \rho\, \mathbf{M}^{(ce)}(\boldsymbol{\Delta})\right)^{-N_r}\right\}_{\boldsymbol{\Delta}}. \tag{3}$$

The expectation is over the set of all error events with uniform probability distribution, and N_e is the cardinality of this set. The $N_t \times N_t$ identity matrix is $\mathbf{I}_{N_t}$.

Self-interference. The squared code matrix is

$$\mathbf{X}^\dagger\mathbf{X} = \tfrac{1}{2}\sum_{k=1}^{2K} c_k^2\, \mathbf{S}^{(kk)} + \sum_{k>l} c_k\, c_l\, \mathbf{S}^{(kl)}, \tag{4}$$

where

$$\mathbf{S}^{(kl)} = \mathbf{B}^{(k)\,\dagger}\mathbf{B}^{(l)} + \mathbf{B}^{(l)\,\dagger}\mathbf{B}^{(k)}. \tag{5}$$

Thus $\mathbf{S}^{(kk)}$ encodes the spatial transmit diversity distribution of symbol c_k, and $\mathbf{S}^{(kl)}$ gives the interference between symbols c_k, c_l. The total self-interference power of $\mathbf{X}$ is given by

$$I = \sum_{k>l} \text{Tr}\left(\mathbf{S}^{(kl)^2}\right). \tag{6}$$

Two symbols with vanishing $\mathbf{S}^{(kl)}$ are Radon-Hurwitz orthogonal (RHO). If all symbols are RHO, the scheme is an (orthogonal) space-time block code with $I = 0$ as discussed by Tarok & al.[5]

Equivalent Channel Matrix. Instead of the matrix signal model (1), a vector signal model may be used. Expressing $\mathbf{Y}$ as a $(TN_r) \times 1$ vector $\mathbf{y}$, the signal model becomes

$$\mathbf{y} = \mathcal{H}\, \mathbf{c} + \text{noise}. \tag{7}$$

The equivalent channel matrix $\mathcal{H}$ is a $TN_r \times 2K$ matrix which depends on the structure of $\mathbf{X}$ and the channel. The (real valued) matched filter outputs corresponding to (7) are

$$\mathbf{z} = \tfrac{1}{2}\left[\mathcal{H}^\dagger\ \mathcal{H}^T\right]\begin{bmatrix}\mathbf{y}\\ \mathbf{y}^*\end{bmatrix} \equiv \mathbf{R}\,\mathbf{c} + \text{noise}, \tag{8}$$

where the $2K \times 2K$ matched filter correlation matrix is $\mathbf{R} = \text{Re}\,\mathcal{H}^\dagger\mathcal{H}$. For a linear matrix modulation of the form (2), the matrix elements of $\mathbf{R}$ are given in terms of traces, and may be expressed in terms of the interference matrices (5) as

$$\mathbf{R}_{jk} = \tfrac{1}{2}\,\text{Tr}\left[\mathbf{S}^{(j,k)}\,\mathbf{H}\,\mathbf{H}^\dagger\right] \tag{9}$$

Mutual Information. If c_k are Gaussian signals with covariance $\mathbf{C} = \frac{TP}{2K}\mathbf{I}_{2K}$, the ergodic mutual information (per channel use) in terms of the correlation matrix $\mathbf{R}$ is

$$\mathcal{I} = \frac{1}{2T}\mathcal{E}\left\{\log\det\left(\mathbf{I}_{2K} + \rho\frac{T}{K}\mathbf{R}\right)\right\}_{\mathbf{H}}. \qquad (10)$$

The correlation matrix $\mathbf{R}$ is normalized so that the signal-to-noise ratio is $\rho = P/\sigma^2$, in terms of the noise variance σ^2 and the average transmit power per channel use $P = \text{Tr}\,\mathbf{C}/T = \mathcal{E}\left\{\text{Tr}\,\mathbf{X}^\dagger\mathbf{X}\right\}/T$. Consistency requires

$$\sum_{k=1}^{2K}\text{Tr}\,\mathbf{S}^{(kk)} = 4K. \qquad (11)$$

Mutual information expansion. Mutual information is generically a non-transparent object. It has been argued that capacity may be approached term by term in low-SNR expansions of the capacity and mutual information[19]; $C = \sum_{n=1}^{\infty}\rho^n C_n$, $\mathcal{I} = \sum_{n=1}^{\infty}\rho^n\mathcal{I}_n$. A capacity reaching modulation scheme has $\mathcal{I}_n = C_n$ for all n. As discussed by Verdu[20], the two first expansion coefficients of $\mathcal{I}$ determine the slope of the spectral efficiency curve (as a function of E_b/N_0) at minimum E_b/N_0.

For i.i.d. Rayleigh fading, the expansion coefficients for the capacity were derived in[19]. The two first coefficients are $C_1 = N_r$ and $C_2 = -N_r(N_r + N_t)/2N_t$.

Mutual information (10) may be expanded using the property $\log\det = \text{Tr}\log$, followed by the expansion $\log x = \sum_{n=1}^{\infty}(-1)^{n+1}x^n/n$. This gives[10]

$$\mathcal{I} = \frac{\rho}{2K}\text{E}\langle\text{Tr}_K\mathbf{R}\rangle_{\mathbf{H}} - \frac{T\rho^2}{4K^2}\text{E}\langle\text{Tr}_K\mathbf{R}^2\rangle_{\mathbf{H}} + \cdots \qquad (12)$$

Here the notation Tr_K stresses that the trace is over symbol indexes. For i.i.d. Rayleigh fading, the channels may be integrated out explicitly. Using (9,11), the linear term in ρ reads

$$\mathcal{I}_1 = \frac{1}{4K}\sum_{n=1}^{N_r}\sum_{k=1}^{2K}\text{Tr}_{N_t}\,\mathbf{S}^{(kk)} = N_r, \qquad (13)$$

Thus first order capacity C_1 is reached if all the power is used.

For linear matrix modulation, the second order information coefficient is always negative[10]:

$$\mathcal{I}_2 = -\frac{TN_r}{16K^2}\sum_{k,l}\left(\text{Tr}\left(\mathbf{S}^{(kl)}\right)^2 + N_r\left(\text{Tr}\,\mathbf{S}^{(kl)}\right)^2\right) \qquad (14)$$

Below we shall argue that maximizing $\mathcal{I}_2$ affects information and performance for all values of SNR.

3. ELEMENTARY SYMMETRIC POLYNOMIALS AND UNITARY INVARIANTS

Unitarily invariant polynomials of matrix elements, or unitary invariants for short, are functions of matrices that are invariant under unitary similarity transformations $\mathbf{M} \mapsto \mathbf{U}^\dagger \mathbf{M} \mathbf{U}$. The most commonly considered unitary invariants are the trace and the determinant of a matrix. The rank of a matrix is also invariant under unitary transformations. We shall see its interpretation in terms of unitary invariants below.

A Hermitian matrix may be diagonalized with a unitary transformation

$$\mathbf{M} = \mathbf{U}^\dagger \Lambda \mathbf{U} , \tag{15}$$

where Λ is a diagonal matrix of eigenvalues λ_i. Thus eigenvalues of a Hermitian matrix are trivially unitary invariants, and they form a complete basis; all unitary invariants can be expressed in terms of the eigenvalues. Another set of unitary invariants are the traces of powers

$$\mathrm{Tr}\,[\mathbf{M}^n] = \sum_{i=1}^{N} \lambda_i^n , \; n = 1, \ldots, N . \tag{16}$$

From cyclicity of trace it follows that they are unitary invariants of any $N \times N$ matrices $\mathbf{M}$. In this paper we are interested in positive semi-definite Hermitian matrices, and we shall use a refined set of invariants, which is directly related to elementary symmetric polynomials of the eigenvalues.

Recall that the N elementary symmetric polynomials E_n of N non-negative variables λ_n are defined as (see e.g.[21])

$$\prod_{n=1}^{N} (1 + \rho \lambda_n) = 1 + \sum_{n=1}^{N} \rho^n E_n(\lambda) , \tag{17}$$

where

$$E_n(\lambda) = \sum_{\sum m\, k_m = n} \frac{(-1)^{n + \sum k_m}}{\prod m^{k_m} k_m!} \prod_{m=1}^{N} \left(\sum_{i=1}^{N} \lambda_i^m \right)^{k_m} \tag{18}$$

The sum is over all sets of integers $\{k_m\}_{m=1}^{N}$ that constitute a partition of n as $\sum_{m=1}^{N} m\, k_m$. For a positive-semidefinite Hermitian $N \times N$ matrix $\mathbf{M}$, the corresponding unitary invariants are the elementary symmetric polynomials of the eigenvalues, and will be denoted by $E_n(\mathbf{M})$. They are traces of matrices of minors of $\mathbf{M}$. We have

$$\det (\mathbf{I} + \rho \mathbf{M}) = 1 + \sum_{n=1}^{N} \rho^n E_n(\mathbf{M}) . \tag{19}$$

For example, $E_1(\mathbf{M}) = \text{Tr}[\mathbf{M}]$, $E_2(\mathbf{M}) = \frac{1}{2}E_1^2 - \frac{1}{2}\text{Tr}\left[\mathbf{M}^2\right]$ and $E_N(\mathbf{M}) = \det[\mathbf{M}]$. The rank has a straight forward interpretation in terms of the elementary symmetric polynomials: $\text{rank } \mathbf{M} = r \Leftrightarrow E_m(\mathbf{M}) = 0 \,\forall\, m > r$.

Here we are interested in the diagonal (or trace) dominance of performance measures. For this, it is worthwhile to separate the trace from the rest of the matrix. For this, denote the traceless part of $\mathbf{M}$ by

$$\widetilde{\mathbf{M}} = \mathbf{M} - \frac{1}{N}\text{Tr}[\mathbf{M}]\,\mathbf{I}_N\,. \tag{20}$$

We denote the normalized trace of $\mathbf{M}$ as

$$T_1(\mathbf{M}) = \frac{1}{N}\text{Tr}[\mathbf{M}]\,, \tag{21}$$

and the elementary symmetric polynomials of the traceless part as

$$T_n(\mathbf{M}) = (-1)^{n+1}\, n\, E_n(\widetilde{\mathbf{M}})\,, n \geq 2\,. \tag{22}$$

For convenience, the normalization has been chosen so that T_n always has a term $\text{Tr}\left(\widetilde{\mathbf{M}}^n\right)$, and accordingly we shall reserve the name "trace invariants" for the T_n. We have

$$T_2 = \text{Tr}\left[\widetilde{\mathbf{M}}^2\right] = \text{Tr}\left[\mathbf{M}^2\right] - N\,T_1^2 \tag{23}$$

The elementary symmetric polynomials in terms of the trace invariants are

$$E_k = \binom{N}{k}T_1^k + \sum_{n=2}^{k}\frac{(-1)^{n-1}}{n}\binom{N-n}{k-n}T_1^{k-n}\,T_n\,. \tag{24}$$

In particular,

$$\det \mathbf{M} = T_1^N + \sum_{n=2}^{N}\frac{(-1)^{n-1}}{n}\,T_1^{N-n}\,T_n\,. \tag{25}$$

From (24) it follows that

$$\frac{dE_k}{dT_n} = \frac{(-1)^{n-1}}{n}\binom{N-n}{k-n}T_1^{k-n}\,,\forall\,k \geq n\,. \tag{26}$$

For a positive definite matrix the trace (and T_1) is positive. Thus the derivative of all E_k, $k \geq n$, and thus of $\det(\mathbf{I} + \rho\mathbf{M})$ with respect to T_n is positive (negative) for n odd (even).

The usefulness of the expansion (25) stems from the fact that Hermitian matrices are diagonal dominated. This is neatly visible in the Hadamard and Schur inequalities, see the Appendix. In particular, the Trace Corollary of

Hadamard Inequality (A.2) tells that the expansion of the determinant (25) is dominated by the first term. The sum of the remaining $N - 1$ terms is always negative, and the absolute value of the sum is at most as big as the first term. The Trace Corollary of Schur Inequality (A.3) tells to what extent the two first terms dominate the expansion. From (A.3), an upper limit for the sum of the remaining terms can be calculated. If $N = 2$, there are only two terms in the expansion (25), and instead of an inequality like (A.3), one just has the equality (25).

4. DIAGONAL DOMINANCE

When designing linear space–time modulations, we are interested in two matrices; the matched filter correlator $\mathbf{R}$, and the distance matrix $\mathbf{M}^{(co)}$, appearing in the mutual information (10) and union bound (3), respectively. The former should be maximized, the latter minimized. In this section we explore these fimctionals in terms of the results of the previous section.

4.1 Diagonal dominance of mutual information

The mutual information (10) is an expectation of the logarithm of (19) with $\mathbf{M} = \mathbf{R}$. $\mathbf{R}$ is positive definite for all non-vanishing channels. Trace-dominance of the mutual information follows from the Trace-corollary of Hadamard inequality (A.2), which gives an upper bound of mutual information. When extremizing this with respect to $\sum_k \mathbf{S}^{(kk)}$ (realizing the power constraint (11 with a Lagrange multiplier), one gets

$$\sum_{k=1}^{2K} \mathbf{S}^{(kk)} \sim \mathbf{I}_{N_t}. \tag{27}$$

This is the tight frame condition found by Heath & al.[6]. The next step in exploring the diagonal dominance considers the effects of the second trace invariant T_2.

Theorem 1: For a linear matrix modulation, the mutual information increases with increasing second order expansion coefficient $\mathcal{I}_2$ at any SNR.

Proof: Using (23) for $\mathbf{M} = \mathbf{R}$, and (13) and (14) for calculating expectation values, we get

$$\mathcal{E}\{T_2(\mathbf{R})\} = \mathcal{E}\left\{ \text{Tr}\left(\mathbf{R}^2\right) - \frac{1}{2K}\left(\text{Tr}\,\mathbf{R}\right)^2 \right\}$$
$$= \frac{N_r}{4}\sum_{k,l}\left(\text{Tr}\left(\mathbf{S}^{(kl)}\right)^2 + N_r\left(\text{Tr}\,\mathbf{S}^{(kl)}\right)^2 - 4N_r\delta_{k,l}\right) \tag{28}$$

Thus $\mathcal{E}\{T_2\}$ equals the second order information (14) up to a negative constant. For X, Y two random variables, $dX/dY > 0$ implies

$$d\mathcal{E}\{X\}/d\mathcal{E}\{Y\} > 0 \tag{29}$$

for the expectation values. From this and (26) it follows directly that

$$\frac{d\mathcal{I}}{d\,\mathcal{E}\{T_2(\mathbf{R})\}} < 0. \tag{30}$$

That is, minimizing $\mathcal{E}\{T_2\}$, i.e. maximizing the second order mutual information $\mathcal{I}_2$, leads to greater mutual information for all SNR values ρ. ∎

4.2 Diagonal dominance of union bound

The union bound of pairwise error probabilities is proportional to an expectation of the **power** $-N_r$th power of (19) with $\mathbf{M} = \mathbf{M}^{(cc)}$. The distance matrix $\mathbf{M}^{(cc)}$ is positive definite for all non-vanishing error events. Trace-dominance of the union bound follows again from (A.2), which gives a lower bound of performance. When extremizing this with respect to $\mathrm{Tr}\,\mathbf{S}^{(kl)}$, $k \neq l$ we find the criterion of Frobenius orthogonality (traceless self-interference)[9],

$$\mathrm{Tr}\,\mathbf{S}^{(kl)} = 0\,, k \neq l. \tag{31}$$

Recalling (5), Equation (31) means that the basis matrices of two real symbols should be orthogonal with respect to the Frobenius norm. The next step is again to consider the second trace invariant T_2.

Theorem 2: For a linear matrix modulation with a Frobenius orthogonal basis, the union bound of pairwise error probabilities increases with increasing self-interference at any SNR.

Proof: As in Theorem 1, it is straight forward consequence of (3) and (26) that

$$\frac{d\,P_{\mathrm{UB}}}{d\,\mathcal{E}\{T_2(\mathbf{M}^{(cc)})\}} > 0. \tag{32}$$

Now in (23) with $\mathbf{M} = \mathbf{M}^{(cc)}$ and a Frobenius orthogonal basis, we have $T_1 = 2KE$. Here (11) was used, and $E = \mathcal{E}\{\Delta^2\}$ is the average value of the squared one-symbol error with uniform distribution in errors, as in the union bound. This can be assumed to be the same for all symbols, as possible differences in symbols can be absorbed into the normalization of basis matrices.

Furthermore,

$$\mathcal{E}\left\{\text{Tr}\left(\mathbf{M}^{(\text{co})2}\right)\right\} = \tfrac{1}{4}\sum_{k,l}\mathcal{E}\left\{\Delta_k^2\Delta_l^2\right\}\text{Tr}\left(\mathbf{S}^{(kk)}\mathbf{S}^{(ll)}\right)$$

$$+E^2\sum_{k>l}\text{Tr}\left(\mathbf{S}^{(kl)2}\right). \qquad (33)$$

The last term is exactly the total self-interference (6) of $\mathbf{X}$, and it is positive semidefinite. The theorem follows from applying the chain rule. ∎

The expression of $\mathcal{E}\left\{T_2\left(\mathbf{M}^{(\text{co})}\right)\right\}$ simplifies if we assume that we assume that all basis matrices are unitary, so that all $\mathbf{S}^{(kk)}$ are proportional to identity with the same proportionality constant. Then we have

$$\mathcal{E}\left\{T_2\left(\mathbf{M}^{(\text{ce})}\right)\right\} = E^2\sum_{k>l}\text{Tr}\left(\mathbf{S}^{(kl)^2}\right), \qquad (34)$$

and $\mathcal{E}\left\{T_2\right\}$ is exactly the self-interference. Thus to minimize the union bound, irrespective of the SNR, the self-interference should be minimized.

Note that also (28) depends directly on the self-interference (6). Both information maximization and performance optimization leads to the same, intuitively clear criterion: the self-interference (non-orthogonality) should be as small as possible[22].

In the theorems above, the implicit assumption is made, that *only* the second order information or the self-interference are changed, keeping all the other degrees of freedom constant. In particular, for Theorem 2 this means that the diversity matrices $\mathbf{S}^{(kk)}$ should be kept constant, along with all higher trace invariants. Given that the trace invariants are highly non-linear functions, finding proper variables to variate in order to minimize the union bound according to (32) is likely to be exceedingly difficult. However, variables that do not change the self-interference I and the second order information $\mathcal{I}_2$ are easy to find. All orthogonal symbol rotations used to optimize performance in the literature[8–12,14] leave I and $\mathcal{I}_2$ invariant but affect the higher trace invariants. Thus Theorem 2 should be understood so that the union bound is separately optimized with respect to I and the symbol rotations, which are independent variables.

5.   EXAMPLE: RATE 3 FOR 4 TX ANTENNAS

When the symbol rate $K/T = N_t$, it is rather straight forward to find information optimal bases. Rate 4 bases of 4×4 matrices have been widely discussed[7,10,12]. Hassibi and Hochwlad[7] proposed full symbol rate $K/T = N_t$

bases of unitary $N_t \times N_t$ matrices for any number of transmit antennas. The underlying algebraic structure of these bases is that of Weyl algebra[14]. Similarly, El Gamal and Damen[12] constructed bases of $N_t \times N_t$ matrices for any N_t by using Hadamard transforms on "threads" (main and side diagonals of the matrix). Furthermore, for orthogonal design dimensions $N_t = 2^m$, $m \in \mathbb{N}$, a Clifford-algebra basis with full symbol rate N_t can be constructed[10].

Lower rate versions of these bases may be constructed by removing symbols. From Clifford bases, layers that constitute an orthogonal design may be removed. Removing complete threads[12] or cyclic layers[14] works for Hadamard and Weyl-bases. All 4 combinations of Hadamard/Weyl bases with threaded/cyclic layering give the same second order information[14]. Thus, from the information point of view, it is sufficient to consider only one of these alternatives. We choose the Weyl basis with cyclic layering,

$$\begin{aligned} \mathbf{X}_{\text{Weyl 3}} &= \mathbf{X}_{\text{cyclic}}(x_1, x_2, x_3, x_4) \\ &+ \mathbf{D}_{\text{FT}}\, \mathbf{X}_{\text{cyclic}}(x_5, \ldots, x_8) + \mathbf{D}_{\text{FT}}^2\, \mathbf{X}_{\text{cyclic}}(x_9, \ldots, x_{12}) \,. \end{aligned} \tag{35}$$

where the cyclic layer is given by

$$\mathbf{X}_{\text{cyclic}}(x_1, x_2, x_3, x_4) = \begin{bmatrix} x_1 & x_2 & x_3 & x_4 \\ x_4 & x_1 & x_2 & x_3 \\ x_3 & x_4 & x_1 & x_2 \\ x_2 & x_3 & x_4 & x_1 \end{bmatrix} \tag{36}$$

with x_k complex symbols, and the layers are separated by the action of the diagonal matrix that generates rows of the 4×4 DFT matrix,

$$\mathbf{D}_{\text{FT}} = \text{diag}[1 \quad j \quad -1 \quad -j] \,. \tag{37}$$

The rate 3 restriction of the Clifford basis with largest $\mathcal{I}_2$ is based on considering the Clifford basis as a collection of orthogonal designs of two symbols, as in the rate 1 ABBA-code[22]. This triple-ABBA is of the form

$$\begin{aligned} \mathbf{X}_{\text{TripABBA}} &= \begin{bmatrix} \mathbf{X}_A + \mathbf{X}_C & \mathbf{X}_B + \mathbf{X}_D \\ j(\mathbf{X}_B - \mathbf{X}_D) & \mathbf{X}_A - \mathbf{X}_C \end{bmatrix} \\ &+ \sqrt{\frac{N_r}{2N_r - 1}} \begin{bmatrix} \mathbf{X}_E + j\mathbf{X}_F & \mathbf{X}_E + j\mathbf{X}_F \\ j(\mathbf{X}_E + j\mathbf{X}_F) & -\mathbf{X}_E - j\mathbf{X}_F \end{bmatrix} , \end{aligned} \tag{38}$$

where $\mathbf{X}_A, \mathbf{X}_B$ etc. are 2×2 Alamouti blocks.

The matrix modulators above are compared to the baseline symbol rate 3 vector modulation ("V-BLAST modulation") with one antenna periodically punctured. For four channel uses, this corresponds to removing one "thread"

Table 19.1. Mutual information properties of symbol rate 3 schemes.

Scheme	$\mathcal{I}_2$	$\mathcal{I}/\mathcal{C}$ (dB) at SNR		
		0 dB	9 dB	15 dB
trip-ABBA	−0.45	−0.01	−0.02	−0.02
Weyl 3	−0.58	−0.21	−0.32	−0.50
baseline	−0.58	−0.21	−0.32	−0.50

(but not applying any transformations on the threads):

$$\mathbf{X}_{\text{baseline}} = \begin{bmatrix} x_1 & x_2 & x_3 & 0 \\ 0 & x_4 & x_5 & x_6 \\ x_7 & 0 & x_8 & x_9 \\ x_{10} & x_{11} & 0 & x_{12} \end{bmatrix}. \tag{39}$$

Information properties of these designs can be found in Table 19.1. Column $\mathcal{I}_2$ gives the second-order information (14) in i.i.d MIMO channels with three Rx antennas. The normalization is (11). The last three columns report the loss from $N_t = 4$, $N_r = 3$ continuous input capacity at three SNR values. The $\mathcal{I}_2$ optimal triple-ABBA scheme almost reaches capacity, whereas the other two are information equivalent, and slightly suboptimal. This behavior is predicted by (30).

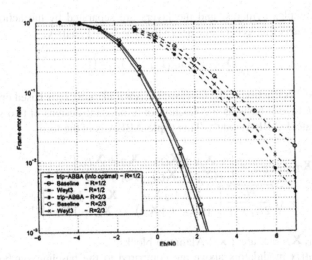

Figure 19.1. Frame error rate of symbol rate 3 modulators concatenated with rate 1/2 convolutional code, 4 Tx, 3 Rx, 3 bps/Hz.

Simulated performance in i.i.d. block Rayleigh fading is reported in Figure 19.1. Instead of high SNR optimization using constellation rotation, performance with a randomly interleaved concatenated outer channel code is investigated. A constraint length 7 rate $R = 1/2$ convolutional code (CC), and the same CC punctured to rate 2/3 using puncturing patterns proposed by Yasuda & al.[23], have been concatenated to the matrix modulators (38,35,39). Iterative detection/decoding is performed, with soft interference cancelation. First X is LMMSE detected, the output is passed through parallel interference cancelation (PIC), followed by max-log MAP decoding of the CC. At subsequent iterations the soft-output from max-log MAP decoder is used in the soft PIC instead of the LMMSE estimates. PIC of X and max-log MAP decoding of CC are iterated six times. Note that the resulting difference in decoding complexity of a matrix vs. vector modulator is rather insignificant.

The simulation confirms the diagonal dominance results of the previous section. Irrespective of the rate, the second order information optimal scheme outperforms the others over the whole SNR range, with a margin of the order of the information gain. Also, it is noticeable that the rate 1/2 outer code is able to exploit almost all of the implicit transmit diversity promised by the channel through the channel code induced redundancy. With rate 1/2, the performance difference of baseline vector modulation with no transmit diversity and Weyl 3 with explicit transmit diversity is insignificant. With a $R = 2/3$ outer code, the matrix modulators (38,35) with explicit diversity show a clear gain at high SNR.

6.   CONCLUSION

We discussed the self-interference and the second trace invariant T_2 of the matrices appearing in information and performance measures for linear matrix modulations, and observed that a systematic design of the basis of space–time matrices gives better performing space–time modulations than ad hoc methods. The work in Section 3 applies equally well to any space–time code. The elementary symmetric polynomials E_k furnish the full spectrum of space–time code design criteria, interpolating between trace (dominant at SNR=0) and determinant (dominant at SNR $= \infty$). The trace invariant T_n affects all $E_k, k \geq n$. Thus the higher the SNR, the more trace invariants need to be optimized.

A.   Determinant Inequalities

The diagonal dominance of positive semidefinite matricdes can be expressed in terms of a number of inequalities.

Theorem A.1: **Hadamard Inequality.** For a positive semidefinite Hermitian matrix, the determinant is upper bounded by the product of the diagonal elements:

$$\det \mathbf{M} \leq \prod_{i=1}^{N} m_{ii} \tag{A.1}$$

A corresponding lower bound is provided by an inequality due to Schur (see e.g.[24]).

Theorem A.2: **Schur Inequality.** For a positive semidefinite Hermitian $N \times N$ matrix with $N > 2$, the determinant is lower bounded by

$$\det \mathbf{M} \geq \prod_{i=1}^{N} m_{ii} - \frac{\mathrm{Tr}[\mathbf{M}]^{N-2}}{2(N-2)} \left(\mathrm{Tr}\left[\mathbf{M}^2\right] - \sum_{i=1}^{N} m_{ii}^2 \right).$$

Positive semidefinite Hermitian matrices are not only "diagonal dominated", they are "trace dominated". Corresponding looser versions of the inequalities above may be written in terms of the trace invariants T_n. Correponding to the Hadamard and Schur Inequalities, we have

Corollary A.1: **Trace Corollary of Hadamard Inequality.** For a positive semi-definite Hermitian $N \times N$ matrix, the determinant is upper bounded by

$$\det \mathbf{M} \leq T_1^N. \tag{A.2}$$

Proof: Follows directly from Theorem A.1 and the geometric-arithmetic mean inequality. ■

Corollary A.2: **Trace Corollary of Schur Inequality.** For a positive semi-definite Hermitian $N \times N$ matrix with $N > 2$, the determinant is lower bounded by

$$\det \mathbf{M} \geq \max \left(0, \; T_1^N - \frac{(NT_1)^{N-2}}{2(N-2)} T_2 \right). \tag{A.3}$$

Proof: Consider the function $f = \prod_{i=1}^{N} m_{ii} + \frac{(NT_1)^{N-2}}{2(N-2)} \sum_{i=1}^{N} m_{ii}^2$ on the constraint surface with a fixed trace, $NT_1 = \sum_{i=1}^{N} m_{ii}$. Using (23), it is sufficient to prove that $f \geq \left(1 + \frac{N^{N-1}}{2(N-2)} \right) T_1^N$. The extrema can be found by differentiating $f + \lambda \left(NT_1 - \sum_{i=1}^{N} m_{ii} \right)$, where λ is a Lagrange multiplier imposing the constraint. For any λ, the only solution for the extremal equations is $m_{ii} = T_1 \; \forall \; i$. On the constraint surface the Hessian of f with respect to the independent variables has off-diagonal elements $\left(\frac{N^{N-2}}{N-2} - 1 \right) T_1^{N-2}$ with diagonal elements twice as big. All eigenvalues of the Hessian are positive. Thus the extremum with all m_{ii} equal is a global minimum at the constraint surface. The value of f at this minimum is $\left(1 + \frac{N^{N-1}}{2(N-2)} \right) T_1^N$. ∎

References

1. G. Foschini, "Layered space–time architecture for wireless communication in a fading environment when using multi–element antennas," *Bell Labs Tech. J.*, pp. 41–59, 1996.

2. E. Telatar, "Capacity of multi-antenna gaussian channels," *Eur. Trans. Telecomm.*, vol. 10, no. 6, pp. 585–595, Nov/Dec 1999.

3. J. Winters, "On the capacity of radio communication systems with diversity in a Rayleigh fading environment," *IEEE J. Sel. Areas Comm.*, vol. 5, no. 5, pp. 871–878, June 1987.

4. P. Wolniansky, G. Foschini, G. Golden, and R. Valenzuela, "V-BLAST: An architecture for realizing very high data rates over the rich-scattering wireless channel," in *Proc. URSI Int. Symp. on Signals, Systems and Electronics*, Sept. 1998, pp. 295–300.

5. V. Tarokh, H. Jafarkhani, and A. Calderbank, "Space–time block codes from orthogonal designs," *IEEE Trans. Inf. Th.*, vol. 45, no. 5, pp. 1456–1467, July 1999.

6. R. Heath, Jr., H. Bölcskei, and A. Paulraj, "Space–time signaling and frame theory," in *Proc. IEEE ICASSP*, 2001, vol. 2, pp. 1194–1199.

7. B. Hassibi and B. Hochwald, "High-rate codes that are linear in space and time," *IEEE Trans. Inf. Th.*, vol. 48, no. 7, pp. 1804–1824, July 2002.

8. M. Damen and N. Beaulieu, "A study of some space–time codes with rates beyond one symbol per channel use," in *Proc. IEEE GLOBECOM*, Nov. 2001, vol. 1, pp. 445–449.

9. O. Tirkkonen and A. Hottinen, "Improved MIMO performance with non-orthogonal space–time block codes," in *Proc. IEEE GLOBECOM*, Nov. 2001, vol. 2, pp. 1122–1126.

10. O. Tirkkonen and R. Kashaev, "Combined information and performance optimization of linear MIMO modulations," in *Proc. IEEE ISIT*, July 2002, p. 76.

11. R. Heath, Jr. and A. Paulraj, "Linear dispersion codes for MIMO systems based on frame theory," *IEEE Trans. Sign. Proc.,* vol. 50, no. 10, pp. 2429–2441, Oct. 2002.

12. H. El Gamal and M. Damen, "Universal space–time coding," *IEEE Trans. Inf. Th.,* 2003.

13. R. Kashaev and O. Tirkkonen, "Linear matrix modulators from group representation theory," in *Proc. IEEE Inf. Th. Worksh.,* Mar. 2003, pp. 42–45.

14. A. Hottinen, O. Tirkkonen, and R. Wichman, *Multiantenna Transceiver Techniques for 3G and Beyond,* Chichester: John Wiley and Sons, 2003.

15. J.-C. Guey, M. Fitz, M. Bell, and W.-Y. Kuo, "Signal design for transmitter diversity wireless communication systems over Rayleigh fading channels," *in Proc. IEEE VTC,* Spring, 1996, pp. 136–140.

16. V. Tarokh, N. Seshadri, and A. Calderbank, "Space–time codes for high data-rate wireless communication: Performance criterion and code construction," *IEEE Trans. Inf. Th.,* vol. 44, no. 2, pp. 744–765, Mar. 1998.

17. D. M. Ionescu, "New results on space–time code design criteria," in *Proc. IEEE WCNC,* Sept. 1999, pp. 684–687.

18. E. Biglieri, G. Taricco, and A. Tulino, "Performance of space–time codes for a large number of antennas," *IEEE Trans. Inf. Th.,* vol. 48, no. 7, pp. 1794–1803, July 2002.

19. R. Kashaev and O. Tirkkonen, "On expansion of MIMO mutual information in SNR," in *Proc. IEEE ISIT,* July 2002, p. 252.

20. S. Verdú, "Spectral efficiency in the wideband regime," *IEEE Trans. Inf. Th.,* vol. 48, no. 2, pp. 1319–1343, June 2002.

21. R. Horn and C. Johnson, *Matrix Analysis,* Cambridge University Press, 1985.

22. O. Tirkkonen, A. Boariu, and A. Hottinen, "Minimal non-orthogonality rate one space–time block code for 3+ Tx antennas," in *Proc. IEEE ISSSTA,* Sept. 2000, vol. 2, pp. 429–432.

23. Y. Yasuda, K. Kashiki, and Y. Hirata, "High-rate punctured convolutional codes for soft decision Viterbi decoding," *IEEE Trans. Comm.,* vol. COM-32, no. 3, pp. 315–319, Mar. 1984.

24. A. Marshall and I. Olkin, *Inequalities: Theory of Majorization and its Applications,* New York: Academic Press, 1979.

CONTRIBUTORS

PART I   TRENDS IN WIRELESS NETWORKS

Chapter 1   **CROSS-LAYER PERFORMANCE IN CELLULAR WCDMA/3G NETWORKS: MODELLING AND ANALYSIS**

EKRAM HOSSAIN is working as an Assistant Professor in the Department of Electrical and Computer Engineering at University of Manitoba, Winnipeg, Canada. He received his Ph.D. in electrical engineering from University of Victoria, Canada, in 2000. He was a University of Victoria Fellow. He was a recipient of the Lucent Technologies, Inc. research award for his contributions to the IEEE International Conference on personal Wireless Communications (ICPWC), 1997. Dr. Hossain's research interests include radio link control and transport layer protocol design and cross-layer optimization issues for the next-generation wireless data networks. He leads the *Wireless Internet and Packet Radio Network Research Group* in the Department of Electrical and Computer Engineering at University of Manitoba. Currently he serves as an Editor for the *IEEE Transactions on Wireless Communications* and the *IEEE/KICS Journal of Communications and Networks*.

VIJAY K. BHARGAVA received the B.Sc., M.Sc., and Ph.D. degrees from Queen's University, Kingston, Canada in 1970, 1972 and 1974, respectively. Currently, he is a Professor and Chair of the Department of Electrical and Computer Engineering at the University of British Columbia. He was also awarded a Tier I Canada Research Chair in 2001, which he held until 2003. He is a co-author of the book *Digital Communications by Satellite* (New York: Wiley, 1981) and co-editor of the IEEE Press Book *Reed-Solomon Codes and Their Applications*. Dr. Bhargava is a Fellow of the B.C. Advanced Systems Institute, Engineering Institute of Canada (EIC), the IEEE, the Canadian Academy of Engineering and the Royal Society of Canada. He is a recipient of the IEEE Centennial Medal (1984), IEEE Canada's McNaughton

Gold Medal (1995), the IEEE Haraden Pratt Award (1999), the IEEE Third Millennium Medal (2000), and the IEEE Graduate Teaching Award (2002).

Chapter 2   ALWAYS ON SERVICE INTELLIGENT NETWORK

SARIT MUKHERJEE is a lead researcher in the Bell Laboratories' Center for Networking Research, leading the research and development of the next generation wireless data service technologies. Before this he held a Technical Manager position with Lucent Technologies managing the research and development of internet content distribution appliances. Prior to this he managed the design and development of streaming appliances in a New York-based start-up company, led the video networking group in Panasonic Information and Networking Technology Lab at Princeton, and taught as an assistant professor of Computer Science and Engineering at the University of Nebraska-Lincoln. He received his Ph.D. in Computer Science from the University of Maryland, College Park in 1993. He published in several renowned technical journals and conferences, served in the technical committees of number of international conferences, and holds dozen of US patents. His research interests include high speed network architectures and protocols, wireless networks and multimedia applications.

SANJOY PAUL is currently the Director of Wireless Networking Research Department at Bell Laboratories. Before that he was the Chief Technology Officer at Edgix. He has over fifteen years of technology expertise, specifically in the areas of multicasting, media streaming, intelligent caching, mobile networking, and secure commerce. Prior to joining Edgix, Sanjoy was a Distinguished Member of Technical Staff at the Bell Laboratories Research, where he was the chief architect and visionary of Lucent's IPWorX (later called Imminet) caching and content distribution product line. Sanjoy is well regarded in the technical community for his contributions to the field of Internetworking: designing the Reliable Multicast Transport Protocol (RMTP), holding twenty U.S patents, publishing a book on Multicasting and numerous papers, and receiving the 1997 William R. Bennett award from IEEE Communications Society for the best original paper published in IEEE/ACM Transactions on Networking. Sanjoy is a Fellow of the IEEE, an editor of IEEE/ACM Transactions on Networking, and an adjunct faculty of WINLAB at Rutgers University. He holds a Bachelor of Technology degree from Indian Institute of Technology, Kharagpur, India and both an M.S and a Ph.D. degree from the University of Maryland, College Park.

KRISHAN SABNANI is currently the Senior Vice President of the Networking Research Laboratory at Bell Labs in New Jersey. He has been a member of Bell Labs Research for the past 21 years. During this time he conceived and launched a number of systems projects in the areas of Internetworking and wireless networking, led successful transfers of research ideas to products in the Lucent and AT&T business units, and conducted extensive personal research in data and wireless networking. Krishan is a Bell Labs Fellow. He is also a fellow of the Institute of

Electrical and Electronic Engineers (IEEE) and of the Association of Computing Machinery (ACM). He received the Leonard G. Abraham award from the IEEE Communications Society in 1991 and the Bell Laboratories Distinguished Technical Staff Award in 1990. He also received the President of India's Gold Medal and the Institution of Engineers (India) Gold Medal, both in 1975. Krishan received his Ph.D. in electrical engineering from Columbia University, New York, in 1981, and joined Bell Labs that same year. He received B. Tech. from IIT Delhi in 1975.

Chapter 3   COGNITIVE TRENDS IN MAKING

PETRI MÄHÖNEN is currently a full professor and holder of Ericsson chair of wireless networks at Aachen University in Germany. Previously he studied and worked in the United States, United Kingdom, and Finland, most recently as research director of networking at the Centre for Wireless Communications, Oulu, Finland. He has been principal investigator in several international research projects, including several large European Union research projects for wireless communications and networking research. Since November 2003 he has been chairman of WG3 (cooperative and adhoc networks) of the World Wireless Research Forum. His current research focuses on wireless Internet, low-power communications including sensors, cognitive networks and radios, broadband wireless access, applied mathematical methods for telecommunications, and theory of co-operative networks.

Chapter 4   MULTIPLE ANTENNA SYSTEMS: FRONTIER OF WIRELESS ACCESS

ENRICO DEL RE [SM] (delre@lenst.det.unifi.it) received a Dr. Ing. degree in electronics engineering from the Universityof Pisa, Italy, in 1971. Until 1975 he was engaged in public administration and private firms, involved in the analysis and design of telecommunication and air traffic control equipment and space systems. Since 1975 he has been with the Department of Electronics Engineering of the University of Florence, Italy, first as a research assistant, then as an associate professor, and since 1986 as a professor. During the academic year 1987–1988 he was on leave from the University of Florence for a nine-month period of research at the European Space Research and Technology Centre of the European Space Agency, The Netherlands. His main research interests are digital signal processing, mobile and satellite communications, and communication networks, on which he has published more than 150 papers in international journals and conferences. He is coeditor of the book *Satellite Integrated Communications Networks* (North-Holland, 1988), and one of the authors of the book *Data Compression and Error Control Techniques with Applications* (Academic, 1985). He has been chairman of European Project COST 227, "Integrated Space/Terrestrial Mobile Networks," and EU COST Action 252, "Evolution of Satellite Personal Communications from Second to Future Generation Systems." He received the 1988–1989 premium from the IEE (UK) for the paper "Multicarrier Demodulator for Digital Satellite Communication

Systems." He is head of the Digital Signal Processing and Telematics Laboratory of the Department of Electronics and Telecommunications of the University of Florence. He is a member of the Executive Board of the Italian Interuniversity Consortium for Telecommunications (CNIT). He is a member of the European Association for Signal Processing (EURASIP).

LAURA PIERUCCI (pierucci@lenst.det.unifi.it) received her Dr. Eng. degree in electronics engineering from the University of Florence, Italy, in 1987. Until 1991 she was a research consultant for private firms in the fields of ultrasonic applications in medicine and the development of radio mobile cellular systems. In 1989 she was winner of a fellowship of the National Research Council (CNR.) on "Study of Devices in Optical Fibre and of Optical Multiplexing Techniques to Use in Data Transmission Network." Currently, she is with the Department of Electronics and Telecommunications of the University of Florence, as assistant professor in the fields of telecommunications and digital signal processing. Her main research interests are in digital signal processing, adaptive systems, neural network, antenna array, MIMO systems, and the general area of mobile and satellite communication systems. Since 1990, she took part in different national projects of MURST and the Italian Space Agency (ASI) and was involved in the European Projects ISIS — Interactive Satellite multimedia Information System (1995–1998), Galenos — Generic Advanced Low Cost trans European Network Over Satellite (1999–2001), Hermes (1998–2002) on the fields of telecommunications systems on terrestrial and satellite networks and their applications such as tele-education and tele-medicine. She was the scientific secretariat of EU COST Action 252, "Evolution of Satellite Personal Communications from Second to Future Generation Systems" (1996–2000).

Chapter 5   FIXED RELAYS FOR NEXT GENERATION WIRELESS SYSTEMS

NORBERT ESSELING received his Diploma in Electrical Engineering in 1994 from the Aachen University of Technology. From 1994 to 1996 he worked at T-Mobil, Bonn/Germany. He was responsible for the D1-ISUP specification and also involved in the development of GSM core net signaling specifications and extensions (e.g. convergence GSM to fixed and satellite). In 1996 he joined the Chair for Communication Networks (ComNets) at Aachen University of Technology, where he was working towards his Ph.D. He participated in the ACTS Project SAMBA (wireless ATM) where he was involved in the design, implementation and integration of the SAMBA trial platform. Areas of research interest at ComNets were protocols to support wireless broadband packet networks. His focus was on aspects extending the range of the HiperLAN/2 System. In 2002 he re-joined T-Mobile International and is currently working in the area of IP based multimedia systems, especially dealing with end-to-end QoS topics.

BERNHARD H. WALKE for the last 13 years is running the Chair for Communication Networks at RWTH Aachen University, Germany, where about 35 researchers work under his guidance on topics like air-interface design, formal specification of protocols, fixed network planning, development of tools for stochastic event driven simulation and analytical performance evaluation of services and protocols of XG wireless systems. During that time he has supervised more than 650 Master theses and 43 Ph.D. theses covering most aspects of fixed and mobile communication networks. This work continuously has been funded from third parties' grants. He has published more than 120 reviewed conference papers, 25 journal papers and seven textbooks on architecture, traffic performance evaluation and design of future communication systems. Prior to joining academia, Prof. Walke worked in various industry positions for AEG-Telefunken (now part of EADS AG). Prof. Walke holds diploma and Dr. degrees in electrical engineering, both from the University of Stuttgart, Germany.

RALF PABST (pab@comnets.rwth-aachen.de) was born in Cologne, Germany in 1974. After having absolved student internships at SIEMENS ICN and D2 Vodafone, he received his Diploma in Electrical Engineering from Aachen University in 2001 and serves since then as a research assistant at the chair of Communication Networks (Prof. B. Walke) of Aachen University, where he is working towards his Ph.D. degree. He has been working on various IST Projects (WSI, WWRI, NEXWAY and ANWIRE) and is actively involved in the organization of the WG4 in WWRF. His research interests include the performance of relay-based deployment concepts for next-generation wireless networks, the associated Resource Management Protocols and the spectral coexistence of wireless standards.

Chapter 6   DYNAMIC ENHANCEMENT AND OPTIMAL UTILIZATION OF CDMA NETWORKS

JOSEPH SHAPIRA is the president of Comm&Sens, a strategic consulting company in wireless communications, sensing and imaging systems. He is the founder and past chairman of Celletra Ltd. – a cellular enhancement solutions company. During his tenure at Qualcomm he took a major part in the development of the CDMA standard. He then founded and was the President of Qualcomm Israel. Before joining Qualcomm, Dr. Shapira held a broad spectrum of research and management staff positions in the largest R&D organization in Israel, including establishment and leadership of the Electromagnetics lab. Dr. Shapira received his B.Sc. and M.Sc. in Electrical Engineering from the Technion - Israel Institute of Technology, and obtained a Ph.D. degree in Electrophysics from the Polytechnic University. He was awarded best paper award from IEEE AP-S in 1974, and the Bergman award by the president of Israel in 1980 for developing Electromagnetics in Israel. He is a Fellow of the IEEE, and served 6 years as a Vice President of URSI – the International Union for Radio Sciences. He is a member of Forum 100 – influential alumni of the Technion – Israel Institute of technology.

PART II   WIRELESS LANS AND AD HOC NETWORKS

Chapter 7   ARCHITECTURE AND PROTOTYPING OF AN 802.11 BASED SELF-ORGANIZING HIERARCHICAL ADHOC WIRELESS NETWORK (SOHAN)

SACHIN GANU received the B.E degree from Mumbai University, Mumbai, India in 1998, and the M.S degree from Virginia Tech, Blacksburg, VA, in 1999. He is working toward the Ph.D. degree in WINLAB, Department of Electrical Engineering, Rutgers University, Piscataway, NJ. He worked as a Member of Technical Staff with Comsat Laboratories, Clarksburg, MD from Feb 2000 until July 2000 and as a Systems Engineer with Lockheed Martin Global Telecommunications, Clarksburg, MD from July 2000 until Dec 2001. He worked as a Research Scientist at Avaya Research Labs, Basking Ridge, NJ during the summer of 2002. His research interests include wireless adhoc networks, sensor networks with focus on interlayer interactions between Physical Layer, MAC and Routing for adhoc networks.

LALIT Y. RAJU received the B.Tech degree from the National Institute of Technology, Hamirpur (H.P.), India in 2001 and the M.S degree from Rutgers University, New Brunswick, NJ in May 2004. His current research interests include topology control and routing protocols for adhoc and sensor networks.

BHASKAR ANEPU received the B.Tech degree from Jawaharlal Nehru Technological University (JNTU), Hyderabad, India in 2001, and the M.S degree from Rutgers University, New Brunswick, NJ, in 2004. His research interests include wireless system design, link layer and network layer protocol development for wireless adhoc networks and application development for wireless networks.

SULI ZHAO received her Master's and bachelor's degrees in Electrical Engineering at Beijing University of Posts and Telecommunications, Beijing, China, in 1997 and 1994, respectively. She is working toward the Ph.D. degree in Electrical and Computer Engineering at Rutgers, The State University of New Jersey, She worked as a research engineer at Nokia Research Center, Beijing, from 1998 to 2001, concentrating on radio access networks for 3G mobile communications. Her current interests focus on mobile wireless networking, hierarchical adhoc network architecture, adhoc network routing protocols, and cross layer design for adhoc networks.

IVAN SESKAR [M'89] (seskar@winlab.rutgers.edu) received a B.S. degree in Electrical Engineering and Computer Science from the University of Novi Sad, Yugoslavia, and an M.S. degree in Electrical Engineering from Rutgers University. Since 1991, he has been at the Wireless Information Networks Laboratory (WINLAB) at Rutgers University, where he is currently the associate director of information technology. His current research interests include wireless adhoc networks, sensor networks, spectrum management, 3G and WLAN systems.

DIPANKAR RAYCHAUDHURI is Professor, Electrical & Computer Engineering Department and Director, WINLAB (Wireless Information Network Lab) at Rutgers University. As WINLAB's Director, he is responsible for a cooperative industry-university research center with focus on next-generation wireless technologies. WINLAB's current research scope includes topics such as RF/sensor devices, UWB, spectrum management, future 3G and WLAN systems, adhoc networks and pervasive computing. He has previously held progressively responsible corporate R&D positions in the telecom/networking area including: Chief Scientist, Iospan Wireless (2000-01), Assistant General Manager & Dept Head-Systems Architecture, NEC USA C&C Research Laboratories (1993-99) and Head, Broadband Communications Research, Sarnoff Corp (1990-92). Dr. Raychaudhuri obtained his B.Tech (Hons) from the Indian Institute of Technology, Kharagpur in 1976 and the M.S. and Ph.D degrees from SUNY, Stony Brook in 1978, 79. He is a Fellow of the IEEE.

Chapter 8   VOICE CAPACITY IN IEEE 802.11 NETWORKS

MONCEF ELAOUD (M'97) received his B. Sc. (1988) his M. Sc. (1990) and his Ph. D. (2000) in electrical and computer engineering from the University of Wisconsin-Madison. He is currently a senior research Scientist at Telcordia Technologies' Applied Research organization. His main research interests are in the areas of quality of service, self-forming and self healing networks, auto-configuration, and mobility management in wireless and adhoc networks. He is currently leading a QoS team to study and develop QoS solutions in Wi-Fi networks. He is also leading an effort to design and implement a new QoS architecture for adhoc networks as part of the Future Combat System program.

PRATHIMA AGRAWAL is the Samuel Ginn Distinguished Professor of Electrical and Computer Engineering at Auburn University, Auburn, Alabama. She is also the Director of the Wireless Engineering Research and Education Center at Auburn. Earlier she was Assistant Vice President of the Network Systems Research Laboratory and Executive Director of the Mobile Networking Research Department at Telcordia Technologies Morristown, NJ, where she has worked since 1998. She was the Head of the Networked Computing Research Department in AT&T/Lucent Bell Laboratories in Murray Hill, NJ, where she worked from 1978 to 1998 in various capacities. Dr. Agrawal's research interests are computer networks, mobile and wireless computing and communication systems. She has published over 150 papers and holds 32 patents. Dr. Agrawal is a Fellow of the IEEE and a Member of the ACM. Dr. Agrawal received her Ph.D. degree in Electrical Engineering from the University of Southern California in 1977.

Chapter 9   AD HOC WIRLESS NETWORKING USING MOBILE BACKBONES

IZHAK RUBIN received the B.Sc. and M.Sc. from the Technion, Israel, and the Ph.D. degree from Princeton University, all in Electrical Engineering. Since 1970, he has been on the faculty of the UCLA School of Engineering and Applied Science where he is a Professor in the Electrical Engineering Department. During 1979-1980, he served as Acting Chief Scientist of the Xerox Telecommunications Network. He served as co-chairman of the 1981 IEEE International Symposium on Information Theory and as program chairman for the 1987 IEEE INFOCOM conference. Dr. Rubin is a Fellow of IEEE. He has served as editor of the IEEE Transactions on Communications, and of the journals: Wireless Networks, Optical Networks magazine, Photonic Network Communications and Communications Systems. Over the last several years, he has been responsible for developing adhoc wireless networking and management architectures and protocols for unmanned-vehicle aided adhoc multi-tier wireless networks.

ARASH BEHZAD received the BSc degree in industrial engineering (system design and analysis) from Azad University, Tehran, Iran, in 1996. He received the MSc degree in industrial engineering with an emphasis on optimization from Sharif University of Technology, Tehran, Iran, in 1998. He served as a lecturer at Azad University in 1999. Since 2000, he has been working toward the PhD degree in electrical engineering at the University of California (UCLA). He was a recipient of the University of California MICRO fellowship during the academic year 2000-2001. His current area of research includes communications and telecommunications systems and networks, applied probability theory, and combinatorial optimization. He is a member of the IEEE, American Mathematical Society (AMS), and the Institute of Operations Research and Management Sciences (INFORMS).

HUEI-JIUN JU received her B.S. degree and M.S. degree from National Taiwan University in 1999 and 2001, respectively, both in Electrical Engineering. She is currently a Ph.D. student in the Electrical Engineering department at UCLA. She joined the Autonomous Intelligent Networked Systems Lab at 2001. Her research focuses on adhoc wireless networking, especially protocols at the MAC, Network and Transport layers. Her recent work includes enhancing UDP and TCP performance in multi-hop adhoc networks using an adaptive IEEE 802.11 MAC layer RTS/CTS engagement algorithm, as well as developing a scalable fully distributed Mobile Backbone Network (MBN) topology synthesis algorithm.

RUNHE ZHANG received the B.S degree from Jiaotong University, Shanghai, China in 1998. After working as technical staff at Lucent Technologies for one and half year, he attended the University of Minnesota, Twin Cities and earned his M.S.E.E in 2001. That same year he joined the University of California, Los Angeles where he is pursuing his doctoral degree in Electrical Engineering.

XIAOLONG HUANG received his Bachelor degree in Electrical Engineering from Peking University, China, in 1999. He received his Master degree in Electrical Engineering from Hong Kong University of Science and Technology, Hong Kong, in 2001. He is currently a Ph.D student in the Electrical Engineering Department of the University of California, Los Angeles, USA. He joined the Motorola R&D Center (China) as an internship engineer for software testing in 1997. In 2003, He joined the HITACHI (China) R&D Center as an internship researcher for protocol design of MIPv6 home agents. His research interests include routing and MAC protocols in wireless adhoc networks, topological synthesis protocols for Mobile Backbone Networks, and Quality of Service in adhoc networks.

YICHEN LIU received the B.Eng degree from Shanghai Jiao Tong University (SJTU) in electrical engineering, Shanghai, China in 1999, and the M.Phil degree in electrical and electronic engineering from the Hong Kong University of Science and Technology (HKUST), Hong Kong, Chinain 2001. He has been a Ph.D student in electrical engineering department at the University of California, Los Angeles since 2001, conducting research under the supervision of Prof. Izhak Rubin. While at SJTU and HKUST, he was a member of the Telecommunication Technology Laboratory and of the High Performance Network Switching Laboratory, respectively, working as a research assistant. Since 2001, he has been working at the Autonomous Intelligent Networked Systems Laboratory. His research interests include Mobile Backbone Network (MBN) topology synthesis, high performance routing and flow control in adhoc wireless networks.

RIMA KHALAF received the B.S. in Engineering from the American University of Beirut (with distinction) in 2001, and the M.S. in Electrical Engineering from UCLA in 2003. She is currently a PhD student at UCLA where she works under the supervision of Prof. I. Rubin. Her research interests are focused on developing MAC and Routing Protocols for Adhoc wireless networks. She is also involved in the implementation of power control MAC algorithms for IEEE802.11 cards.

Chapter 10   VERTEX-LINKED INFRASTRUCTURE FOR AD HOC NETWORKS

VICTOR ON-KWOK LI received SB, SM, EE and ScD degrees in Electrical Engineering and Computer Science from the Massachusetts Institute of Technology in 1977, 1979, 1980, and 1981, respectively. He is Chair Professor of Information Engineering at the University of Hong Kong. Previously, he was Professor of Electrical Engineering at the University of Southern California (USC), Los Angeles, California, USA, and Director of the USC Communication Sciences Institute. Sought by government, industry, and academic organizations, he has lectured and consulted extensively around the world. His research interest is in information technologies, focusing on the Internet and wireless networks. He has received numerous awards, including the Outstanding Researcher Award of the University of Hong Kong, the KC Wong Education Foundation Lectureship, the Croucher

Foundation Senior Research Fellowship, and the Bronze Bauhinia Star, Government of the Hong Kong Special Administrative Region, China. He is a Fellow of the IEEE, the IAE, and the HKIE.

Chapter 11   PROBABILISTIC METHODS FOR LOCATION ESTIMATION IN WIRELESS NETWORKS

PETRI T. KONTKANEN received his MSc degree from the University of Helsinki in 2004 and is currently working as a researcher in the Complex Systems Computation Group. His research interests are in Bayesian and information-theoretic modeling. He has been a program committee member for the UAI'03 and UAI'04 conferences, and he has published over 40 papers in books, journals and conferences in the areas of probabilistic and information-theoretic modeling, data analysis and visualization, machine learning and stochastic optimization.

PETRI MYLLYMÄKI obtained his MSc degree in Computer Science at University of Helsinki in 1991 and his PhD in 1995, and he became a Decent in computer science in 1999. Dr. Myllymäki is one of the co-founders of the Complex Systems Computation (CoSCo) research group, and he has over 15 years of experience in research in the area of intelligent systems. He has been an editorial board member, program committee member and reviewer for several international scientific journals and conferences, and he has published over 70 scientific articles in his research area. He has also worked as a project manager in numerous applied research projects, and the co-operation with the industry has lead to a number of fielded applications and patents. Dr. Myllymäki is currently working as a professor at the Department of Computer Science of University of Helsinki.

TEEMU T. ROOS received the MSc degree from the University of Helsinki in 2001. He is currently pursuing a PhD in computer science. He has co-authored theoretical papers on machine learning and Bayesian modeling and applied papers on location estimation. Since 1999, he has been with the Complex Systems Computation Group (Helsinki Institute for Information Technology). His current research interests are primarily in Bayesian and information-theoretic data analysis, probabilistic graphical models, and machine learning.

HENRY R. TIRRI received his BSc, MSc and PhD in Computer Science from University of Helsinki, Finland. He is currently a Nokia Research Fellow. Dr. Tirri's academic experience includes both research and teaching positions at the University of Helsinki, University of Texas at Austin, MCC, AT&T Bell Laboratories, Purdue University, NASA Ames Research Center, Stanford University, and UC Berkeley. From 1998 he has been a Professor of Computer Science at the University of Helsinki, and an Adjunct Professor of Computational Engineering at the Helsinki University of Technology. He was the Research Director of Intelligent Systems area at Helsinki Institute for Information Technology (2000-2004). He has published more than 160 international publications in books, journals and conferences in the areas of probabilistic and information-theoretic modeling, neural networks, text and data mining, case-based reasoning, transaction processing, intelligent learning environments. His work is both theoretical and applied in nature, and the applied work has a strong multidisciplinary flavor.

KIMMO A. VALTONEN studied computer science and computational linguistics at the University of Helsinki. He has been a member of the Complex Systems Computation group since 1997. He has worked mostly on applications of Bayesian networks and information theory in diverse domains such as the prediction of wild salmon production, the modelling of pharmaceutical data, and location estimation.

HANNES WETTIG received his MSc degree in Mathematics from the University of Köln in 1998. Before that he visited the Complex Systems research group (CoSCo) in Helsinki for the academic year 1996/97 financed by a scholarship issued by the Finnish Center for International Mobility (CIMO). He returned to CoSCO in 1999 as a PhD student of computer science. His current special research interests include Bayesian and information-theoretic data modeling, in particular discriminative learning and Minimum Description Length (MDL) coding.

PART III   MOBILE WIRELESS INTERNET AND SATELLITE APPLICATIONS

Chapter 12   COPING WITH UNCERTAINTY IN MOBILE WIRELESS NETWORKS

SAJAL K DAS is a Professor of Computer Science and Engineering and also the Founding Director of the Center for Research in Wireless Mobility and Networking (CReWMaN) at the University of Texas at Arlington (UTA). His current research interests include resource and mobility management in wireless networks, mobile and pervasive computing, wireless multimedia and QoS provisioning, sensor networks, mobile Internet protocols, distributed processing and grid computing. He has published over 250 research papers, directed numerous funded projects, and holds 5 US patents in wireless mobile networks. He received the Best Paper Awards in ACM MobiCom'99, ICOIN'01, ACM MSWIM'00, and ACM/IEEE PADS'97. Dr. Das is also a recipeint of UTA's Outstanding Faculty Research Award in Computer Science in 2001 and 2003, and UTA's College of Engineering Excellence in Research Award in 2003. He serves on the Editorial Boards of IEEE Transactions on Mobile Computing, ACM/Kluwer Wireless Networks, Journal of Parallel and Distributed Computing, Parallel Processing Letters, Journal of Parallel Algorithms and Applications. He served as General Chair of IEEE PerCom'04, IWDC'04, CIT'03 and IEEE MASCOTS'02, ACM WoWMoM'00-02; General Vice Chair of PerCom'03, ACM MobiCom'00 and HiPC'00-01; Program Chair of IWDC'02, WoWMoM'98-99; TPC Vice Chair of ICPADS'02, IEEE ICC'03; and as TPC member of numerous IEEE and ACM conferences. He is the Vice Chair of IEEE TCPP and TCCC Executive Committees and on the Advisory Boards of several companies. Prior to 1999, Dr. Das was a professor of computer science at Univeristy of North Texas where he twice (1991 and 1997) received the Student Association's Honor Professor Award for best teaching and scholarly research. He received B.Tech. degree in 1983 from Calcutta University, M.S. degree in 1984 from Indian Institute of Science, Bangalore, and PhD degree in 1988 from the University of Central Florida, Orlando, all in Computer Science.

CHRISTOPHER ROSE received the B.S. (1979), M.S. (1981) and Ph.D. (1985) degrees all from the Massachusetts Institute of Technology in Cambridge, Massachusetts. Following graduate school, Dr. Rose joined AT&T Bell Laboratories in Holmdel, N.J. as a member of the Network Systems Research Department. Dr. Rose is currently a Professor of Electrical & Computer Engineering at Rutgers University in New Jersey, an Associate Director of the Wireless Networks Laboratory (WINLAB), as well as a Henry Rutgers Research Fellow. He is an editor for the ACM Wireless Networks (WINET) Journal, the Elsevier Computer Networks Journal, and has served on many conferences' technical program committees and was technical program co-chair for MobiCom'97, Co-chair of the WINLAB Focus'98 on the U-NII, the WINLAB Berkeley Focus'99 on Radio Networks for Everything and the Berkeley WINLAB Focus 2000 on Picoradio Networks. Dr. Rose, a past member of the ACM SIGMobile Executive Committee is currently a member of the ACM MobiCom Steering Committee and has also served as General Chair of ACM SIGMobile MobiCom 2001 (Rome, July 2001). In December 1999 and 2003 he served on an international panel to evaluate engineering teaching and research in Portugal. Closer to home Dr. Rose has served on the Scientific Fields Advisory Committee of the New Jersey Commission on Science and Technology. His current technical interests include mobility management, novel mobile communications networks, applications of genetic algorithms to control problems in communications networks and most recently, interference avoidance methods using universal radios to foster peaceful coexistence in what will be the wireless ecology of the recently allocated 5GHz U-NII bands.

Chapter 13   ROAMING IN THE GLOBAL WIRELESS INTERNET

CHIP ELLIOTT - As Principal Engineer for BBN Technologies, Mr. Elliott has led the design and successful implementation of a number of secure, mission-critical networks based on novel Internet technology for the United States, Canada, and the U.K., and has acted as senior advisor on a number of national and commercial networks including three LEO satellite constellations and Boeing's Connexion system. Mr. Elliott has particular expertise in wireless Internet technology, mobile "adhoc" networks, quality of service issues, and novel routing techniques. At present he is leading the design and build-out of a very highly secure network protected by quantum cryptography. He holds over 125 patents pending or issued on network technology, currently serves on the Naval Studies Board (National Academy of Sciences) and the Technology Experts Panel for Quantum Cryptography (ARDA), and has participated in a variety of other national advisory panels.

Chapter 14   PROXY SERVICES FOR THE MOBILE INTERNET

VICTOR C.M. LEUNG received the Ph.D. degree in electrical engineering from the University of British Columbia (UBC) in 1981. From 1981 to 1987, Dr. Leung was a Senior Member of Technical Staff at MPR Teltech Ltd., specializing in the planning, design and analysis of satellite communication systems. In 1988, he was a Lecturer in the Department of Electronics at the Chinese University of Hong Kong. Since 1989, he has been a faculty member at the Department of Electrical and Computer Engineering at UBC, where he is a Professor and holder of the TELUS Mobility Industrial Research Chair in Advanced Telecommunications Engineering. His research interests include the design of architectures, protocols and management techniques and the performance analysis of mobile and wireless networks. Dr. Leung is a Fellow of IEEE and a voting member of ACM. He is an editor of the IEEE Transactions on Wireless Communications, and an associate editor of the IEEE Transactions on Vehicular Technology.

Chapter 15   DEVELOPMENT AND FUTURE APPLICATIONS OF SATELLITE COMMUNICATIONS

ERICH LUTZ received the Dipl.-Ing. degree from the Technical University Munich in 1977 and the Dr.-Ing. degree from the University of the Armed Forces, Munich in 1983. Since then, he has been with the Institute of Communications and Navigation of the German Aerospace Center (DLR) and participated in a number of international projects. Since 1986, he has been head of the Digital Network section of this institute. His research interests include networking aspects in mobile and broadband satellite communication systems. Dr. Lutz has published numerous journal and conference papers, and is the principal author of a book on satellite systems for personal and broadband communications published by Springer in 2000. He holds a honorary professorship at the Technical University Munich where he lectures on satellite communication networks. Dr. Lutz is Senior Member of IEEE and member of the Editorial Panel of the International Journal of Satellite Communications.

HERMANN BISCHL received the Dipl.-Ing. degree in electrical engineering from the Technical University Munich in 1989, and the Dr.-Ing. degree from the University of the Federal Armed Forces, Munich, in 1994. Since 1989 he has been with the Digital Networks section of the Institute of Communications and Navigation of the German Aerospace Center (DLR). In 1997 he received an innovations award from the Society of Friends of DLR and in 2001 the DGLR Lectureship Award in the area space on the occasion of the German Aerospace Congress 2001. He led more than 5 and participated in more than 15 national and international research projects. He is reviewer of various journals and conferences. His current research interests include broadband and mobile satellite communications networks, communication protocols, medium access control, error control, channel modeling, DVB-S2/RCS, ATM, IP, MPLS, resource management, and Quality of Service (QoS).

HARALD ERNST received the Dipl.-Ing. degree in electrical engineering in 1996 from the University Karlsruhe, Germany. Until 1998 he worked at the Technical University Munich in the Institute for Communications Engineering in cooperation with DLR, the German Aerospace Center. 1998 he joined DLR as a full time researcher, specializing in satellite communication. He is editor of the ASMS-TF R&D report on mobile usages of satellites. His research interests are satellite based turbo-interference cancellation and efficient satellite multicast approaches for the mobile channel. He is reviewer for different journals, e.g. IEEE Trans. on Communications, IEEE Trans. on Vehicular Technology and International Journal on Satellite Communications and was reviewer in the 5th EU framework program for IST/mobile applications.

FLORIAN DAVID received a Dipl.-Ing. degree in EE from the Technical University of Munich in 1998 and a degree for post-graduate studies in Economics from the Distant Teaching University Hagen in 2002. He submitted his PhD thesis to the Christian-Albrechts-University of Kiel in 2004. He is with the Institute of Communications and Navigation of the German Aerospace Center (DLR) since 1998. His main research interests are optical communications links for space- and airborne platforms, receiver structures and optical beam propagation through the atmosphere.

MATTHIAS HOLZBOCK received the Diplom-Ingenieur (Dipl.-Ing. univ.) degree in electrical engineering and information technology in 1996 from the Technical University of Munich, Germany. From 1995 to 1996 he was with the Optical Networks Group of the German Aerospace Center (DLR). Since 1996 he is with the Institute of Communications and Navigation of the German Aerospace Center (DLR) as a research scientist and project manager. He is Visiting Research Fellow at the University of Bradford since 2001 and also managing partner of TriaGnoSys GmbH. He is member of IEEE and gave lectures at several universities. He authored and co-authored numerous scientific publications and patents.

AXEL JAHN received the PhD and the Diplom-Ingenieur degree in electrical engineering in 1999 and 1990 from the Fern University at Hagen and University Fridericana at Karlsruhe, Germany, respectively. Since September 1990 he is with the Institute of Communications and Navigation of the German Aerospace Center (DLR) as a research scientist, project manager and group leader. He is also a managing partner of TriaGnoSys GmbH. Dr Jahn is a lecturer at the Carl-Cranz-Gesellschaft for Mobile Satellite Communications since 1993 and in 1995/1996 he was a lecturer at Technical University at Ilmenau. Dr Jahn is a Senior Member of the IEEE. He received the DLR Science Award 2002, the Best Paper Award of ITG conference Mobile Communications in 1993 and is listed in Who is Who in Science. He authored and co-authored more than 90 publications, with two scientific textbooks and 13 scientific journal papers.

MARKUS WERNER received the Dipl.-Ing. degree from Darmstadt Technical University, Germany, and the Ph.D. degree from Munich Technical University, Germany, both in electrical engineering. Since 1991, he has been with the Institute of Communications and Navigation of the German Aerospace Center (DLR), Oberpfaffenhofen, Germany, as research scientist, project manager and group leader. Since 2002, he is also a managing partner of TriaGnoSys GmbH, Wessling, Germany, a consulting company for satellite and aeronautical communications. His project experience includes several national and ESA studies and various projects in the framework of European ACTS, IST, and COST research programmes. His main research activities are in the areas of multiservice traffic engineering and capacity dimensioning for satellite systems in general, and system design issues for aeronautical satellite systems in particular. He is co-author of the textbook *Satellite Systems for Personal and Broadband Communications* (Berlin, Germany: Springer-Verlag, 2000).

Chapter 16   ROLE OF SATELLITES IN MOBILE/WIRELESS SYSTEMS

BARRY G EVANS was educated at the University of Leeds, obtaining BSc (1st Class Hons.) and PhD. degrees in 1965/8 respectively. He then joined the British Telecom sponsored team at the University of Essex (Lecturer-Reader 1968-83) where he was responsible for Telecommunications Systems post-graduate activities and radio and satellite research. In 1983 he was appointed to the Alex Harley Reeves Chair of Information Systems Engineering at the University of Surrey. In 1990, he was appointed Director of the Centre for Satellite Engineering Research and in 1996 to the Directorship of the Centre for Communication Systems Research (CCSR) which he continues to hold. He was Dean of Engineering from 1998-2001 and was then appointed Pro-Vice-Chancellor for Research and Enterprise which he continues to hold. He has served on UK Foresight, MoD-DSAC and ETU Committees and is currently on the Ofcom Special Advisory Committee. He is editor of the International Journal of Satellite Communications and the author of three books on telecommunications and satellites as well as over 500 papers in learned journals. He was elected to a Fellowship of the Royal Academy of Engineering in 1991.

PART IV   ENCODING, ALGORITHMS AND PERFORMANCE

Chapter 17   APPLYING NEAR SHANNON-LIMIT CODES TO WIRELESS COMMUNICATIONS

MUSTAFA EROZ received his B.S. from Bilkent University, Ankara, Turkey in 1992 and Ph.D. from the University of Maryland in 1996, both in Electrical

Engineering. Since 1996, he has been with the Advanced Development Group of Hughes Network Systems (HNS), Germantown, MD. His current research interests include low density parity check (LDPC) codes, parallel and serial concatenated codes, product codes, trellis-coded modulation, iterative decoding algorithms and coding for multiple antennas. He discovered a class of semi-regular LDPC codes with near Shannon limit performance and simple implementation which has been standardized by Digital Video Broadcast (DVB). He also played a leading role in specifying the turbo codes for the 3rd generation wireless standards. He received many Hughes Electronics Awards, including the 2002 Chairman's Award, 2003 Hyland Patent Award, 1998, 2001 and 2002 Patent Awarsd, as well as seven patents. Dr. Eroz taught at University of Maryland and Johns Hopkins University.

LIN-NAN LEE received his B.S. degree from National Taiwan University, his M.S. and Ph.D. from the University of Norte Dame, all in Electrical Engineering, in 1970, 1972, and 1976, respectively. Before joining Hughes Network Systems in 1992, he worked at Linkabit Corporation and Communications Satellite Corporation (COMSAT) in many research, development and management capacities. At Hughes, he heads the Advanced Development Group. His research interest included conditional access, digital video coding, channel coding, modulation, multiple access and networking. He and his group have made many significant contributions to the standardization activities within the satellite and wireless industries, as well as design and engineering of HNS satellite and wireless communications products. In recognition of his technical leadership, he was awarded the 2003 Hughes Electronics Chairman's Award. At COMSAT, he received the COMSAT Exceptional Invention Award, and the 1985 and 1988 COMSAT Research Award. Dr. Lee is a Fellow of IEEE.

FENG-WEN SUN is a technical director responsible for R&D on signal processing/coding and modem design. He has extensive publications on error-correction coding, signal processing and synchronization techniques. He has played the leading role in defining and implementing the key algorithms and VLSI architectures for many commercial products. He played a leading role in the adoption of Turbo code and LDPC respectively by third generation wireless and DVB standards. The intellectual property developed by him and his team during these processes has generated millions of dollars license fees. Dr. Sun was the recipient of the 1998 and 2001 Hughes Electronics Patent Award and the co-recipient of the 1998 CDMA technical achievement award. In 2000, he received a special award for "exceptional contributions to third generation wireless technologies". In 2002, he was awarded the Hughes Electronics Chairman's Award. In 2003, He was awarded the P. A. Hyland Patent Award.

Chapter 18   TIME-DIFFUSION CONCEPTS AND PROTOCOL FOR SENSOR NETWORKS

WEILIAN SU received his BS degree in Electrical and Computer Engineering from Rensselaer Polytechnic Institute in 1997. He also received his MS and PhD degree in Electrical and Computer Engineering from Georgia Institute of Technology in 2001 and 2004. Currently, he is an Assistant Professor with the Electrical and Computer Engineering Department, Naval Postgraduate School. In 2003, he received the IEEE Communications Society 2003 Best Tutorial Paper Award. His current research interests include sensor networks, peer-to-peer networks, and wireless networks.

IAN F. AKYILDIZ received his BS, MS, and PhD degrees in Computer Engineering from the University of Erlangen-Nuernberg, Germany, in 1978, 1981 and 1984, respectively. Currently, he is the Ken Byers Distinguished Chair Professor with the School of Electrical and Computer Engineering, Georgia Institute of Technology and Director of Broadband and Wireless Networking Laboratory. He is an Editor-in-Chief of Computer Networks (Elsevier) and for the newly launched Adhoc Networks (Elsevier) Journal. Dr. Akyildiz is an IEEE FELLOW (1995), an ACM FELLOW (1996). He served as a National Lecturer for ACM from 1989 until 1998 and received the ACM Outstanding Distinguished Lecturer Award for 1994. Dr. Akyildiz received the 1997 IEEE Leonard G. Abraham Prize award (IEEE Communications Society) for his paper entitled "Multimedia Group Synchronization Protocols for Integrated Services Architectures" published in the IEEE Journal of Selected Areas in Communications (JSAC) in January 1996. Dr. Akyildiz received the 2002 IEEE Harry M. Goode Memorial award (IEEE Computer Society) with the citation "for significant and pioneering contributions to advanced architectures and protocols for wireless and satellite networking". Dr. Akyildiz received the 2003 IEEE Best Tutorial Award (IEEE Communicaton Society) for his paper entitled "A Survey on Sensor Networks", published in IEEE Communication Magazine, in August 2002. Dr. Akyildiz received the 2003 ACM SIGMOBILE award for his significant contributions to mobile computing and wireless networking. His current research interests are in Sensor Networks, InterPlaNetary Internet, Wireless Networks and Satellite Networks.

Chapter 19   INTERFERENCE INFORMATION AND PERFORMANCE IN LINEAR MATRIX MODULATION

OLAV TIRKKONEN received his M.Sc. and Ph.D. degrees in Theoretical Physics from Helsinki University of Technology in 1990 and 1994, respectively. Since 1999 he is with the Radio Communications laboratory at Nokia Research Center, Helsinki, Finland, where he acts as a principal scientist. At NRC his research has concentrated on baseband algorithms for multiantenna channels, especially space-time coding and MIMO communications. Before joining Nokia, he was with the University of Helsinki, the University of British Columbia, Vancouver, and the Nordic Institute for Theoretical Physics, Copenhagen. He has published 16 journal papers and 22 conference papers, and is a coauthor of the book "Multiantenna transceiver techniques for 3G and beyond".

MIKKO KOKKONEN received M.Sc. and Lic. Tech degrees in Computer Science from Helsinki University of Technology in 1991 and 1999, respectively. In 1993 he joined Nokia Research Center, Helsinki, Finland, where he has been working in the area of digital signal processing methods for radio communications.

ABOUT THE EDITORS

RAJAMANI GANESH received his BE degree from the Indian Institute of Science, Bangalore in 1985. Later, he received his MS and PhD degrees in Wireless Communications from Worcester Polytechnic Institute, Massachusetts (USA) in 1987 and 1991, respectively. He has more than 13 years' work experience in corporate USA where he held many technical management and leadership positions and successfully executed projects of all kinds and sizes. Presently, he works as the Director of Technology Development and Marketing for Qualcomm International with current posting in India. He is actively involved in helping CDMA operators with new technology roadmaps, network performance engineering and deployment of key product offerings. Before joining Qualcomm, he was the Chief Engineer in a mobile-Location based services start-up, which he helped found in the Boston area. Previous to that, he was a Senior Scientist in Verizon Technology Organization (USA) for about 7 years, working on CDMA Network planning, deployments and optimization with special emphasis on capacity enhancement coupled with infrastructure-cost minimization. Earlier, from 1991 to 1995, he was with Sarnoff Corporation in Princeton, New Jersey (USA), working on HDTV transmission and packet CDMA systems. Dr. Ganesh has published dozens of technical papers in many journals and conferences and has received many awards including the prestigious WARNER award, Verizon/GTE's highest award for outstanding technical achievement. He has 15 patents (7 pending) in wireless network planning and optimization issues, mobile Location determination systems & Bluetooth networks. He has contributed to chapters in books and has also edited 2 books on wireless communications: *"Wireless Multemedia Network Technologies"* and *"Wireless Network Deployments"*. He has been the Technical Program Co-Chairman for the last two International IEEE conferences, held in India, on Personal Wireless Communications. He is also active, in the technical program and organizing committees of various other conferences worldwide.

SASTRI L KOTA received his B.S in Physics from Andhra University, B.S.E.E from Birla Institute of Technology and Science (BITS), Pilani and M.S.E.E. from Indian Institute of Technology (IIT) Roorkee, India. He received the Electrical Engineer's Degree from Northeastern University, Boston, MA, U.S.A and Ph.D in Electrical and Information Engineering from University of Oulu, Finland. He is a Senior Scientist at Harris Corporation. Prior to that he held various technical and management positions and contributed to military and commercial communication systems in broadband network architectures and protocols, satellite communication systems design, wireless networks, and performance modeling and analysis at Loral Skynet, Lockheed Martin, SRI International, Ford Aerospace, The MITRE Corp, Xerox Corp, and Computer Sciences Corp. He was on the faculty of Electronics and Communication Engineering Department of Indian Institute of Technology (IIT), Roorkee. He is an active participant at various Standardization Organizations and currently he is the US chair for ITU-R, Working Party 4B and International Rapporteur for Ka-Band Fixed Satellite Systems. He was the chair for Wireless ATM Working Group and has been an ATM Forum Ambassador. His research interests include wireless mobile Adhoc networks, satellite IP networks, QoS and traffic management, broadband satellite access, and ATM networks. He is the principal author of *"Broadband Satellite Communications for Internet Access"*, Kluwer Academic Publishers, 2004 and has contributed book chapters to *"Encyclopedia of Telecommunications"*, John Wiley &Sons, 2003, *"High Performance TCP/IP Networking"*, Prentice Hall, 2003 and *"Modeling and Simulation Environment for Terrestrial and Satellite Networks"*, Kluwer Academic Publishers, 2002. He has published and presented over 100 technical papers in journals, and conference proceedings. He served as a guest editor for special issues for IEEE Communications Magazine, on *Satellite ATM architectures* and *Broadband Network Performance* and International Journal of Satellite Communications and Networking, on *Satellite IP QoS*. He currently serves on the editorial boards of International Journal of Satellite Communications and Networking (Wiley Interscience), and International Journal of Space Communications (IOS Press). He also served as technical chair, member of organizing committees and technical program committees of numerous IEEE, AIAA, SPIE and ACM conferences and workshops. He is the recipient of the Golden Quill Award from Harris Corporation and the ATM Forum Spotlight award. He is a senior member of IEEE, Associate Fellow of AIAA, and Member of ACM.

KAVEH PAHLAVAN is a Professor of ECE, a Professor of CS, and Director of the Center for Wireless Information Network Studies, Worcester Polytechnic Intitute, Worcester, MA. He is also a visiting Professor of Telecommunication Laboratory and CWC, University of Oulu, Finland. His area of research is location aware broadband wireless indoor networks. He has contributed to numerous seminal technical publications in this field. He is the principal author of the Wireless Information Networks (with Allen Levesque), John Wiley and Sons, 1995 and Principles of Wireless Networks – A Unified Approach (with P. Krishnamurthy), Prentice Hall, 2002. He has been a consultant to a number of companies. Before joining WPI, he was the director of advanced development at Infinite Inc., Andover, Mass. working on data communications. He started his career as an assistant

Professor at Northeastern University, Boston, MA. He is the Editor-in-Chief of the International Journal on Wireless Information Networks. He was the founder, the program chairman and organizer of the IEEE Wireless LAN Workshop, Worcester, in 1991 and 1996 and the organizer and technical program chairman of the IEEE International Symposium on Personal, Indoor, and Mobile Radio Communications, Boston, MA, 1992 and 1998. For his contributions to the wireless networks he was the Westin Hadden Professor of Electrical and Computer Engineering at WPI during 1993-1996, was elected as a fellow of the IEEE in 1996 and become a fellow of Nokia in 1999. From May of December of 2000 he was the first Fulbright-Nokia scholar at the University of Oulu, Finland.

Index